SpringerBriefs in Applied Sciences and Technology

For further volumes:
http://www.springer.com/series/8884

Francisco Chinesta · Roland Keunings
Adrien Leygue

The Proper Generalized Decomposition for Advanced Numerical Simulations

A Primer

Springer

Francisco Chinesta
Adrien Leygue
GeM UMR CNRS
Ecole Centrale de Nantes
Nantes
France

Roland Keunings
Applied Mathematics
Université catholique de Louvain
Louvain-la-Neuve
Belgium

ISSN 2191-530X   ISSN 2191-5318   (electronic)
ISBN 978-3-319-02864-4   ISBN 978-3-319-02865-1   (eBook)
DOI 10.1007/978-3-319-02865-1
Springer Cham Heidelberg New York Dordrecht London

Library of Congress Control Number: 2013951633

Printed on acid-free paper

Springer is part of Springer Science+Business Media (www.springer.com)

To Ofelia, Françoise and Samantha

Preface

In spite of the impressive progress made during the last decades in mathematical modelling and techniques of scientific computing, many problems in science and engineering remain intractable. Among them, we can cite those related to high-dimensional models (i.e., in dynamics of complex fluids, quantum chemistry), for which classical mesh-based approaches fail due to the exponential increase of the number of degrees of freedom. Other challenging scenarios are those requiring many direct solutions of a given problem (i.e., in optimization, inverse identification, uncertainty quantification) or those needing very fast solutions (i.e., for real-time simulation, simulation-based control).

Over the last 10 years, the authors of the present book and their collaborators have been developing a novel technique, called Proper Generalized Decomposition (PGD), that has proven capable of tackling these challenging issues. The PGD builds, by means of a successive enrichment strategy, a numerical approximation of the unknown fields in a separated form involving a *priori* unknown functions of individual or clustered coordinates of the problem. Although first introduced and successfully demonstrated in the context of high-dimensional problems, the PGD allows for a completely new approach for addressing more standard problems in science and engineering. Indeed, many challenging problems can be efficiently cast into a multidimensional framework, thus opening entirely new solution strategies in the PGD framework. For instance, the material parameters and boundary conditions appearing in a particular mathematical model can be regarded as extra-coordinates of the problem in addition to the usual coordinates such as space and time. In the PGD framework, this enriched model is solved only once to yield a general or parametric solution that includes all particular solutions for specific values of the parameters. Thus, optimization of complex problems, uncertainty quantification, simulation-based control, and real-time simulation become feasible, even in highly complex scenarios. Once the PGD-separated representation of the parametric solution has been computed offline, its online use only requires one to particularize the solution for a desired set of parameter values. Obviously, this task can be performed very fast and repeatedly in real time, by using light computing platforms such as smartphones or tablets.

The PGD is a numerical technique based on separated representations. The use of separated representations is not new. Indeed, the reader probably remembers the *method of separation of variables* discussed in his/her introductory course on partial differential equations. The basic idea behind this analytical procedure is to assume that the exact solution $u(x, y)$, say of Laplace's equation in a square domain, can be written as an infinite sum of terms, each one consisting of a function of x multiplied by a function of y. When feasible, the method of separation of variables often yields accurate solutions with only a small number of terms. It is unfortunately unable to address nonlinearities or complex geometries, so that the student is quickly introduced to the powerful and flexible alternative of *mesh-based numerical techniques*. Here, an approximation of the solution is computed at a finite number of mesh points distributed in the computational domain, by means of an appropriate technique (e.g. finite elements or finite differences). Suitable interpolation is then used to evaluate the solution anywhere in the computational domain. Powerful numerical methods have been developed to a significant extent over the last decades in virtually all fields of science and engineering. The concurrent explosive increase in computing resources has made possible the numerical simulation of problems of great complexity.

Mesh-based numerical techniques are not feasible, however, for solving mathematical problems defined in spaces of high dimension. The reason is that their number of degrees of freedom grows exponentially with the dimensionality of the problem. High-dimensional models abound in many fields, such as quantum mechanics or kinetic theory of complex materials. In vast contrast, the complexity of the PGD grows linearly with the problem's dimension.

Numerical schemes based on separated representations have been used for decades in the scientific community, under different forms. In quantum chemistry, Hartree-Fock and post-Hartree-Fock approaches exploit this formalism. In the 1980s, P. Ladevèze proposed a space-time separated representation to develop a nonincremental solver for transient problems in computational solid mechanics. It is in the particular context of high-dimensional kinetic models of polymeric fluids that A. Ammar, F. Chinesta, and R. Keunings devised the first version of the PGD strategy in the early 2000s. During the last decade, these authors and other colleagues further developed the PGD and applied it successfully to a wide variety of problems.

The PGD has now attracted the attention of a large number of research groups worldwide. The number of research papers on this topic is growing strongly, and so is the need for an introductory book. The present text is thus the first available book describing the PGD. Our aim has been to provide a very readable and practical introduction, i.e., a "primer", that will allow the reader to quickly grasp the main features of the method. Prerequisites are limited to a basic course in mathematical modeling and numerical methods. Throughout the book, we show how to apply the PGD to problems of increasing complexity, and we illustrate the methodology by means of carefully selected numerical examples. In addition, the reader will have free access to the Matlab software used to generate these examples at http://extras.springer.com. Finally, we provide references to a large number of recent research publications on the PGD that will allow the reader to go beyond the present introduction.

It is a pleasure to acknowledge all our coauthors of PGD-related research publications. In particular, we wish to thank Amine Ammar, Elias Cueto, Antonio Huerta, and Pierre Ladevèze, who from the very beginning contributed significantly to the development of the PGD.

August 2013   Francisco Chinesta
                                                                                 Roland Keunings
                                                                                  Adrien Leygue

Contents

Chapter 1
Introduction

Abstract This chapter summarizes several recurrent issues related to efficient numerical simulations of problems encountered in engineering sciences. In order to alleviate such issues, model reduction techniques constitute an appealing alternative to standard discretization techniques. First, reduction techniques based on Proper Orthogonal Decompositions are revisited. Their use is illustrated and discussed, suggesting the interest of a priori separated representations which are at the heart of the Proper Generalized Decomposition (PGD). The main ideas behind the PGD are described, underlying its potential for addressing standard computational mechanics models in a non-standard way, within a new computational engineering paradigm. The chapter ends with a brief overview of some recent PGD applications in different areas, proving the potentiality of this novel technique.

Keywords Inverse analysis · Model Order Reduction · Multidimensional models · Optimization, Parametric solutions · Proper Generalized Decomposition · Proper Orthogonal Decomposition · Real time · Separated representations · Shape optimization · Simulation based control · Simulation Based Engineering · Virtual charts

1.1 Challenges in Numerical Simulation

In spite of the impressive progress made during the last decades in modelling, numerical analysis, discretization techniques and computer science, many problems in science and engineering remain intractable due to their numerical complexity or to particular constraints such as real-time processing on deployed platforms. We can enumerate different challenging scenarios for efficient numerical simulations:

- The first one concerns models that are defined in high-dimensional spaces. Such models usually appear in quantum chemistry for describing the intimate structure and behavior of materials [1–4]. High-dimensional models also arise, among many other examples, in kinetic theories of complex fluids [5], social dynamics and

F. Chinesta et al., *The Proper Generalized Decomposition for Advanced Numerical Simulations*, SpringerBriefs in Applied Sciences and Technology, DOI: 10.1007/978-3-319-02865-1_1, © The Author(s) 2014

economic systems, vehicular traffic flow phenomena, complex biological systems involving mutation and immune competition, crowds and swarms encountered in congested and panic flows (see [6] and the references therein). A final example is the chemical modelling of systems so dilute that the concept of concentration cannot be used, yielding the so-called chemical master equation that governs cell signalling and other phenomena in molecular biology [7]. Models defined in high-dimensional spaces suffer from the so-called *curse of dimensionality*. If one proceeds with the solution of a model defined in a space of dimension D using a standard mesh-based discretization technique, wherein M nodes are used for discretizing each space coordinate, the resulting number of nodes reaches the astronomical value of M^D. With $M = 10^3$ (a very coarse description in practice) and $D = 30$ (a very simple model) the numerical complexity is 10^{90}, which is larger than the presumed number of elementary particles in the universe! Traditionally, high-dimensional models have been addressed by means of stochastic simulations. These techniques, however, have their own challenges: variance reduction is always an issue and the construction of distribution functions in high-dimensional spaces remains in most cases unaffordable. It is also quite difficult within the stochastic framework to implement parametric or sensitivity analyses that go beyond the brute force approach of computing a large number of expensive, individual simulations.

- Online control of a complex system can be carried out following various approaches. Most commonly one considers the system as a black box whose behavior is modelled by a transfer function relating the system's inputs to its outputs. This pragmatic approach to modelling may seem rather crude but its simplicity is conducive to fast computations. It allows for process control and optimization once a suitable transfer function is available. The tricky point is indeed to obtain such goal-oriented transfer function. For this purpose, it is possible to proceed from a sometimes overly-simplified physical model or directly from experiments, or else from a well-balanced mixture of both approaches. In all cases, the resulting modelling can only be applied within the framework that served to derive it. When a more precise description of the system is needed, however, the simple transfer function modelling approach becomes inapplicable and a detailed physical model of the system is called for. Physical models of complex systems generally are formulated as a set of non-linear and strongly coupled partial differential equations. These rather formidable mathematical objects can indeed be handled using advanced numerical simulation techniques, but this usually requires vast amounts of computing resources. While a detailed numerical simulation yields a wealth of information on the system under investigation, it seems unsuitable in control applications that require fast responses, often in real-time mode. To date, detailed numerical simulation is mostly used offline, but it can in some cases be exploited to formulate simplified models (with their inherent limitations and drawbacks) that can be implemented in real-time mode for online control.

- Many problems in parametric modelling, inverse identification, and process or shape optimization, usually require, when approached with standard techniques, the direct computation of a very large number of solutions of the concerned model for particular values of the problem parameters. When the number of parameters increases such a procedure becomes inefficient.

- Traditionally, the field of simulation-based engineering sciences has relied on the use of static input data to perform the simulations. Input data include parameters of the model(s) or boundary conditions, and the adjective *static* implies that they cannot be modified during the simulation. A new paradigm has emerged in the last decade. Dynamic Data-Driven Application Systems (DDDAS) constitute nowadays one of the most challenging applications of simulations. By DDDAS we mean a set of techniques that allow the linkage of simulation tools with measurement devices for real-time control of complex systems. DDDAS entails the ability to dynamically incorporate additional data into an executing application, and in reverse, the ability of an application to dynamically steer the measurement process [8–10]. In this context, real-time simulators are needed in many applications. One of the most challenging situations is that of haptic devices, where forces must be transferred to the peripheral device at a rate of 500 Hz. Control, identification of malfunction and reconfiguration of malfunctioning systems also need to run in real-time. All these problems can be seen as typical examples of DDDAS.

- Augmented reality is another area in which efficient (i.e. fast and accurate) simulation is urgently needed. The idea is to supply in real-time appropriate information to the reality perceived by the user. Augmented reality could be an excellent tool in many branches of science and engineering. In this context, light computing platforms such as smartphones or tablets are appealing alternatives to large-scale computing facilities.

- Inevitable uncertainty. In science and engineering, in their widest sense, it now seems obvious that there are many causes of variability. The introduction of such variability, randomness and uncertainty in advanced simulation tools is a priority for the next decade.

While the previous list is by no means exhaustive, it includes a set of apparently unrelated problems that can however be treated in a unified manner as will be shown in what follows. Their common feature is that they cannot be solved using direct, traditional numerical techniques. The authors of the present book have contributed actively over the last decade to the development of a new generation of numerical simulation strategies, known as *Proper Generalized Decomposition* (PGD), that has proven highly suitable in addressing these challenges.

The PGD can be viewed as an *a priori* approach to reduced-order modelling. In this chapter, we thus start by illustrating the construction of a reduced-order model using the standard *a posteriori* method of Proper Orthogonal Decomposition (POD). We then introduce the PGD at a glance, and illustrate its capabilities in a wide variety of applications.

1.2 Model Reduction: Information Versus Relevant Information

Consider a discretization mesh having M nodes, and associate to each node an approximation function (e.g. a shape function in the framework of the finite element method). We thus implicitly define an approximation space wherein a discrete solution of the problem is sought. For a transient problem, one must compute at each time step M values (e.g. the nodal values in the finite element framework). For non-linear problems, this usually implies the solution of at least one linear algebraic system of size M at each time step, which becomes computationally expensive when M increases.

In many cases, however, the problem solution lives in a subspace of dimension much smaller than M, and it makes sense to look for a reduced-order model whose solution is computationally much cheaper to obtain. This constitutes the main idea behind the Proper Orthogonal Decomposition (POD), that we now briefly revisit.

1.2.1 Extracting Relevant Information: The Proper Orthogonal Decomposition

We assume that a numerical approximation of the unknown field of interest $u(\mathbf{x}, t)$ is known at the nodes $\mathbf{x}_i$ of a spatial mesh for discrete times $t_m = m \cdot \Delta t$, with $i \in [1, \ldots, M]$ and $m \in [0, \ldots, P]$. We use the notation $u(\mathbf{x}_i, t_m) \equiv u^m(\mathbf{x}_i) \equiv u_i^m$ and define $\mathbf{u}^m$ as the vector of nodal values u_i^m at time t_m. The main objective of the POD is to obtain the most typical or characteristic structure $\phi(\mathbf{x})$ among these $u^m(\mathbf{x})$, $\forall m$ [11]. For this purpose, we maximize the scalar quantity

$$\alpha = \frac{\sum_{m=1}^{P} \left[\sum_{i=1}^{M} \phi(\mathbf{x}_i) u^m(\mathbf{x}_i) \right]^2}{\sum_{i=1}^{M} (\phi(\mathbf{x}_i))^2}, \tag{1.1}$$

which amounts to solving the following eigenvalue problem:

$$\mathbf{c}\boldsymbol{\phi} = \alpha \boldsymbol{\phi}. \tag{1.2}$$

Here, the vector $\boldsymbol{\phi}$ has i-component $\phi(\mathbf{x}_i)$, and $\mathbf{c}$ is the two-point correlation matrix

$$c_{ij} = \sum_{m=1}^{P} u^m(\mathbf{x}_i) u^m(\mathbf{x}_j); \ \mathbf{c} = \sum_{m=1}^{P} \mathbf{u}^m \cdot (\mathbf{u}^m)^T, \tag{1.3}$$

which is symmetric and positive definite. With the matrix $\mathbf{Q}$ defined as

$$\mathbf{Q} = \begin{pmatrix} u_1^1 & u_1^2 & \cdots & u_1^P \\ u_2^1 & u_2^2 & \cdots & u_2^P \\ \vdots & \vdots & \ddots & \vdots \\ u_M^1 & u_M^2 & \cdots & u_M^P \end{pmatrix}, \tag{1.4}$$

we have

$$\mathbf{c} = \mathbf{Q} \cdot \mathbf{Q}^T. \tag{1.5}$$

1.2.2 Building the POD Reduced-Order Model

In order to obtain a reduced-order model, we first solve the eigenvalue problem (1.2) and select the N eigenvectors $\boldsymbol{\phi}_i$ associated with the eigenvalues belonging to the interval defined by the highest eigenvalue α_1 and α_1 divided by a large enough number (e.g. 10^8). In practice, N is found to be much lower than M. These N eigenfunctions $\boldsymbol{\phi}_i$ are then used to approximate the solution $u^m(\mathbf{x})$, $\forall m$. To this end, let us define the matrix $\mathbf{B} = [\boldsymbol{\phi}_1, \ldots, \boldsymbol{\phi}_N]$, i.e.

$$\mathbf{B} = \begin{pmatrix} \phi_1(\mathbf{x}_1) & \phi_2(\mathbf{x}_1) & \cdots & \phi_N(\mathbf{x}_1) \\ \phi_1(\mathbf{x}_2) & \phi_2(\mathbf{x}_2) & \cdots & \phi_N(\mathbf{x}_2) \\ \vdots & \vdots & \ddots & \vdots \\ \phi_1(\mathbf{x}_M) & \phi_2(\mathbf{x}_M) & \cdots & \phi_N(\mathbf{x}_M) \end{pmatrix}. \tag{1.6}$$

Now, we assume for illustrative purposes that an explicit time-stepping scheme is used to compute the discrete solution $\mathbf{u}^{m+1}$ at time t_{m+1}. One must thus solve a linear algebraic system of the form

$$\mathbf{G}^m \, \mathbf{u}^{m+1} = \mathbf{H}^m. \tag{1.7}$$

A reduced-order model is then obtained by approximating $\mathbf{u}^{m+1}$ in the subspace defined by the N eigenvectors $\boldsymbol{\phi}_i$, i.e.

$$\mathbf{u}^{m+1} \approx \sum_{i=1}^{N} \boldsymbol{\phi}_i \, \zeta_i^{m+1} = \mathbf{B} \, \zeta^{m+1}. \tag{1.8}$$

Equation (1.7) then reads

$$\mathbf{G}^m \, \mathbf{B} \, \zeta^{m+1} = \mathbf{H}^m, \tag{1.9}$$

or equivalently

$$\mathbf{B}^T \mathbf{G}^m \, \mathbf{B} \, \zeta^{m+1} = \mathbf{B}^T \mathbf{H}^m. \tag{1.10}$$

The coefficients ζ^{m+1} defining the solution of the reduced-order model are thus obtained by solving an algebraic system of size N instead of M. When $N \ll M$, as is the case in numerous applications, the solution of (1.10) is thus preferred because of its much reduced size.

It is crucial to remark that the reduced-order model (1.10) is built *a posteriori* by means of the already-computed discrete field evolution. Thus, one could wonder about the interest of the whole exercice. In fact, two beneficial approaches are widely considered (see e.g. [12–18]). The first approach consists in solving the large original model over a short time interval, thus allowing for the extraction of the characteristic structure that defines the reduced model. The latter is then solved over larger time intervals, with the associated computing time savings. The other approach consists in solving the original model over the entire time interval, and then using the corresponding reduced model to efficiently solve similar problems with, for example, slight variations in material parameters or boundary conditions.

1.2.3 Illustrating the Construction of a Reduced-Order Model

We consider the following one-dimensional heat transfer problem, written in dimensionless form:

$$\frac{\partial u}{\partial t} = k \frac{\partial^2 u}{\partial x^2}, \tag{1.11}$$

with constant thermal diffusivity $k = 0.01$, $t \in (0, 30]$ and $x \in (0, 1)$. The initial condition is $u(x, t = 0) = 0$ and the boundary conditions are given by $\frac{\partial u}{\partial x}\big|_{x=0,t} = q(t)$ and $\frac{\partial u}{\partial x}\big|_{x=1,t} = 0$.

Equation (1.11) is discretized by using an implicit finite difference method on a mesh with $M = 51$ nodes. The time step is set to $\Delta t = 1$. The resulting discrete system can be written as follows:

$$\mathbf{K}\,\mathbf{u}^{m+1} = \mathbf{M}\,\mathbf{u}^m + \mathbf{q}^{m+1}, \tag{1.12}$$

where the vector $\mathbf{q}^{m+1}$ accounts for the boundary heat flux source at t_{m+1}.

First, we consider the following boundary heat flux source:

$$q(t) = \begin{cases} -1 & t \le 10 \\ 0 & t > 10 \end{cases}. \tag{1.13}$$

The computed temperature profiles are depicted in Fig.1.1 at discrete times $t_m = m$, for $m = 0, 5, 10, 15, 20, 30$. The plain curves correspond to the heating stage up to $t = 10$, while the dashed curves for $t > 10$ illustrate the heat transfer by conduction from the warmest zones towards the coldest ones.

From these 30 discrete temperature profiles, we compute the matrices $\mathbf{Q}$ and $\mathbf{c}$ in order to build the eigenvalue problem (1.2). The 3 largest eigenvalues are found

Fig. 1.1 Temperature profiles corresponding to the source term (1.13) at discrete times $t_m = m$, for $m = 0, 5, 10, 15, 20, 30$. The plain curves correspond to the heating stage up to $t = 10$, while the dashed curves for $t > 10$ illustrate the heat transfer by conduction from the warm zones towards the cold ones

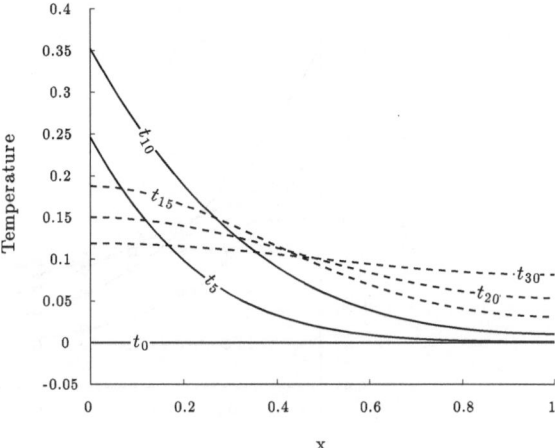

Fig. 1.2 Reduced-order approximation basis involving the eigenvectors corresponding to the three largest eigenvalues

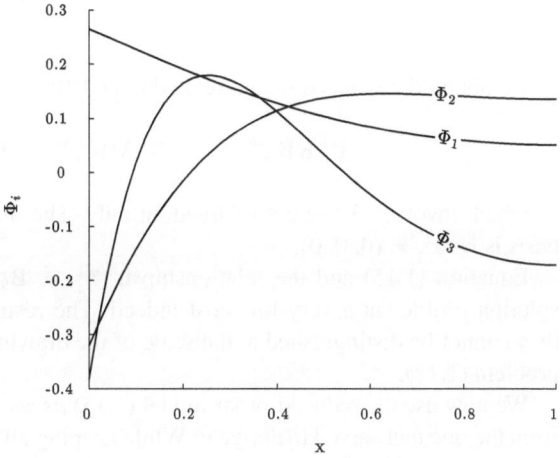

to be $\alpha_1 = 15.537$, $\alpha_2 = 1.178$, $\alpha_3 = 0.076$, while the remaining eigenvalues are such that $\alpha_j < \alpha_1 \cdot 10^{-3}$, $4 \leq j \leq 51$. A reduced-order model involving a linear combination of the 3 eigenvectors related to the first 3 largest eigenvalues should thus be able to approximate the solution with great accuracy. Figure 1.2 shows the resulting approximation functions in normalized form, i.e. $\frac{\phi_j}{\|\phi_j\|}$ ($j = 1, 2, 3$). Defining the matrix **B** as follows:

$$\mathbf{B} = \left[\frac{\phi_1}{\|\phi_1\|} \frac{\phi_2}{\|\phi_2\|} \frac{\phi_3}{\|\phi_3\|} \right], \tag{1.14}$$

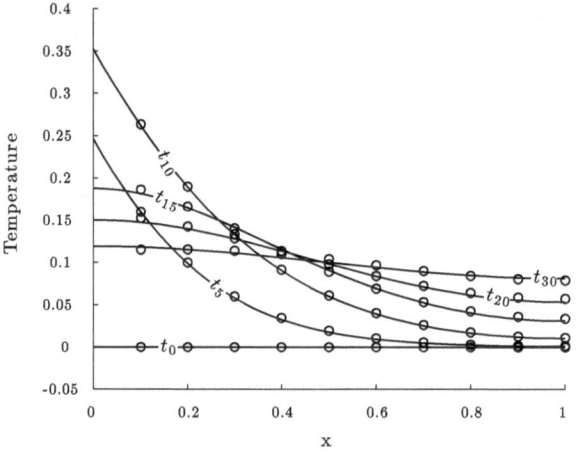

Fig. 1.3 Global (*plain line*) versus reduced-order (*symbols*) model solutions

we obtain the reduced model related to (1.12),

$$\mathbf{B}^T \mathbf{K} \mathbf{B} \, \zeta^{m+1} \; = \; \mathbf{B}^T \mathbf{M} \mathbf{B} \, \zeta^m \; + \; \mathbf{B}^T \mathbf{q}^{m+1}, \tag{1.15}$$

which involves 3 degrees of freedom only. The initial condition in the reduced basis is $(\zeta^0)^T = (0, 0, 0)$.

Equation (1.15) and the relationship $\mathbf{u}^{m+1} = \mathbf{B}\zeta^{m+1}$ then yield approximate solution profiles at a very low cost indeed. The results are shown in Fig. 1.3 and they cannot be distinguished at the scale of the drawing from those of the complete problem (1.12).

We now use the reduced-order model (1.15) *as such* to solve a problem *different* from the one that served to derive it. While keeping all other specifications identical, we now impose instead of (1.13) a substantially different boundary heat flux source:

$$q(t) \; = \; \begin{cases} -\frac{t}{20} & t \le 20 \\[2mm] \frac{30-t}{5} & t > 20 \end{cases}. \tag{1.16}$$

The solution of the reduced model is compared to that of the complete problem in Fig. 1.4. Even though the reduced approximation basis functions are those obtained from the thermal model related to the boundary condition (1.13), the reduced model yields a very accurate representation of the solution of this rather different problem.

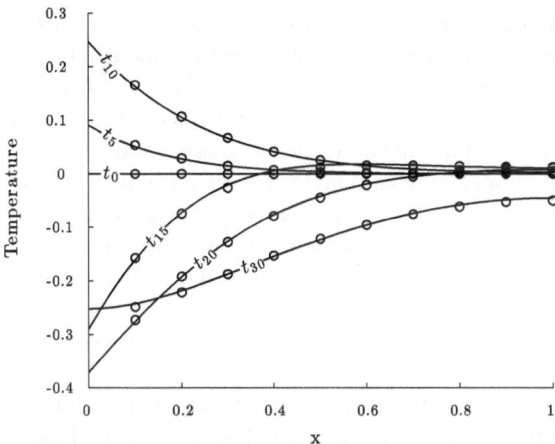

Fig. 1.4 Global (*plain line*) versus reduced-order (*symbols*) model solutions for the source term (1.16). The reduced-order approximation basis is that obtained from the solution of a different thermal problem, with the source term (1.13)

1.2.4 Discussion

The above example illustrates the significant value of model reduction. Of course, one would ideally want to be able to build a reduced-order approximation *a priori*, i.e. without relying on the knowledge of the (approximate) solution of the complete problem. One would then want to be able to assess the accuracy of the reduced-order solution and, if necessary, to enrich the reduced approximation basis in order to improve accuracy. The Proper Generalized Decomposition (PGD), which we describe in general terms in the next section, is an efficient answer to these questions.

The above POD results also tell us that an accurate approximate solution can often be written as a separated representation involving but few terms. Indeed, when the field evolves smoothly, the magnitude of the (ordered) eigenvalues α_i decreases very fast with increasing index i, and the evolution of the field can be approximated from a reduced number of modes. Thus, if we define a cutoff value ϵ (e.g. in the previous example, $\epsilon = 10^{-3} \cdot \alpha_1$, α_1 being the highest eigenvalue), only a small number N of modes are retained ($N \ll M$) such that $\alpha_i \geq \epsilon$, for $i \leq N$, and $\alpha_i < \epsilon$, for $i > N$. Thus, one can write

$$u(\mathbf{x}, t) \approx \sum_{i=1}^{N} \phi_i(\mathbf{x}) \cdot T_i(t) \equiv \sum_{i=1}^{N} X_i(\mathbf{x}) \cdot T_i(t). \tag{1.17}$$

For the sake of clarity, the space modes $\phi_i(\mathbf{x})$ will be denoted in the sequel as $X_i(\mathbf{x})$. Equation (1.17) represents a natural *separated representation*, also known as finite sum decomposition. The solution that depends on space and time can be approximated as a sum of a *small number* of functional products, with one of the functions depending on the space coordinates and the other one on time. Use of separated representations like (1.17) is at the heart of the PGD.

Thus, we expect that the transient solution of numerous problems of interest can be expressed using a significantly reduced number of functional products each involving a function of time and a function of space. Ideally, the functions involved in these products should be determined simultaneously by applying an appropriate algorithm to guarantee robustness and optimality. In view of the non-linear nature of the separated representation, this will require a suitable iterative process.

To our knowledge, the unique precedent to the PGD algorithm for building a separated space-time representation is the so-called radial approximation introduced and intensively developed by Ladevèze ([19–23]) in the context of Computational Solid Mechanics.

In terms of performance, the verdict is simply impressive. Consider a typical transient problem defined in 3D physical space. Use of a standard incremental strategy with P time steps (P is of order of millions in industrial applications) requires the solution of P three-dimensional problems. By contrast, using the space-time separated representation (1.17), we must solve $N \cdot m$ three-dimensional problems for computing the space functions $X_i(\mathbf{x})$, and $N \cdot m$ one-dimensional problems for computing the time functions $T_i(t)$. Here, m is the number of non-linear iterations needed for computing each term of the finite sum (1.17). For many problems of practical interest, we find that $N \cdot m$ is of order 100. The computing time savings afforded by the separated representation can thus reach many orders of magnitude.

1.3 The Proper Generalized Decomposition at a Glance

Consider a problem defined in a space of dimension D for the unknown field $u(x_1, \ldots, x_D)$. Here, the coordinates x_i denote any usual coordinate related to physical space, time, or conformation space, for example, but they could also include problem parameters such as boundary conditions or material parameters as described later. In the simplest approach, we seek a solution for $(x_1, \ldots, x_D) \in \Omega_1 \times \cdots \times \Omega_D$. The PGD yields an approximate solution u^N in the separated form

$$u^N(x_1, \ldots, x_D) = \sum_{i=1}^{N} F_i^1(x_1) \times \cdots \times F_i^D(x_D), \qquad (1.18)$$

where both the number of terms N and the functions $F_i^j(x_j)$ are unknown *a priori*. In this section, and more generally in this book, the approximate solution u^N is denoted by u for notational simplicity.

The representation (1.18) is similar to the classical separation of variables used for solving partial differential equations analytically (in the few instances when this is at all possible!). It is quite general in the sense that any polynomial can be expressed in that form. The main difference between (1.18) and a classical polynomial representation is that the functions $F_i^j(x_j)$, and not only their respective weights, are unknown *a priori*. The price to pay is the non-linear character of the representation (1.18), but the gain is clearly significant as we now discuss. Consider indeed the solution of an hypothetical 2D model given by $u(x, y) = x^n \cdot y^m$. Using a standard Lagrange polynomial approximation, we should consider all monomials up to degree $n + m$,

$$u(x, y) = \sum_{i=0}^{n+m} \sum_{j=0}^{i} \alpha_{i-j, j} \cdot x^{i-j} \cdot y^j. \tag{1.19}$$

Thus, the standard polynomial approximation would involve $\frac{(n+m)\cdot(n+m+1)}{2}$ terms, implying the calculation of their associated weights $\alpha_{i-j, j}$ even though there is only one non-zero coefficient to be obtained in the end, i.e. $\alpha_{n, m} = 1$. To the contrary, the separated representation

$$u(x, y) = \sum_{i=0}^{N} F_i^x(x) \cdot F_i^y(y), \tag{1.20}$$

would capture the solution with only the first term, i.e. $N = 1$, $F_1^x(x) = x^n$ and $F_1^y(y) = y^m$.

The PGD approximation (1.18) is thus a sum of N functional products involving each a number D of functions $F_i^j(x_j)$ that are unknown *a priori*. It is constructed by successive enrichment, whereby each functional product is determined in sequence. At a particular enrichment step $n+1$, the functions $F_i^j(x_j)$ are known for $i \leq n$ from the previous steps, and one must compute the new product involving the D unknown functions $F_{n+1}^j(x_j)$. This is achieved by invoking the weak form of the problem under consideration. The resulting discrete system is non-linear, which implies that iterations are needed at each enrichment step. A one-dimensional problem can thus be defined in Ω_j for each of the D functions $F_{n+1}^j(x_j)$.

If M nodes are used to discretize each coordinate space Ω_j, the total number of PGD unknowns is $N \times M \times D$ instead of the M^D degrees of freedom involved in standard mesh-based discretizations. Moreover, all numerical experiments carried out to date with the PGD show that the number of terms N required to obtain an accurate solution is *not* a function of the problem dimension D, but it rather depends on the separable character of the exact solution. The PGD thus often avoids the exponential complexity with respect to the problem dimension.

In many applications studied to date, the number of terms N in the PGD separated representation (1.18) is found to be as small as a few tens. Furthermore, in all cases referred to in this book, the approximation converges towards the solution associated with the complete tensor product of the approximation bases considered in each Ω_j. Thus, we can be confident about the generality of the separated representation (1.18), but its constructor optimality depends on the nature of the differential operators in the governing equations and on the solution separability. When an exact solution of a particular problem can be represented with enough accuracy by a small number of functional products, the separated approximation is optimal. If the solution is a strictly non-separable function, i.e. a function that requires all the approximation functions generated by the full tensor product of the basis related to each coordinate, the PGD solver proceeds to enrich the approximation until including all the elements of the functional space, i.e. the M^D functions involved in the full tensor product of the approximation bases in each Ω_j.

In the previous pages, we motivated the PGD separated representation from POD-based model reduction. It is important, however, to emphasize the major differences between both approaches.

On one hand, the standard POD is an *a posteriori* approach to model reduction, as it computes at least once an approximate solution of the studied problem using the complete approximation basis. It is from this complete solution that a reduced-order model is extracted as previously described. This reduced model is then used for solving similar problems, and even if these calculations can be performed very fast since the number of degrees of freedom involved in the discrete model scales with the number of functions involved in the reduced basis, one must accept an inevitable error: the reduced basis indeed cannot in general capture all the details related to the solutions of models different from the one from which the reduced basis has been extracted.

On the other hand, the PGD is a radically different, *a priori* approach which does not rely on the availability of a prior solution of the problem. First and foremost, the PGD is an efficient solver that can be applied for calculating the solution of many problems up to the desired accuracy. The errors involved in the computed solution depend on the meshes used for representing the different functions $F_i^j(x_j)$ and on the number of terms N of the finite sum (1.18). Obviously, we could decide to compute only the first few terms of the sum (1.18) and then use them for approximating the solution of similar problems. The PGD could then be viewed as a particularly efficient *a priori* model reduction strategy. In the present book, we discuss the PGD more as an efficient solver than as a reduced modelling technique.

1.4 Towards a New Paradigm in Computational Science

Multidimensional models encountered in kinetic theory of complex materials or in quantum chemistry may seem too far from everyday practice of computational science. More usual models, however, could be enriched with the addition of well chosen extra-coordinates, thus leading to brand new insights into the physics of the problem. Imagine for instance solving the heat diffusion equation with the material thermal conductivity being unknown. This could happen because the conductivity has a stochastic nature or simply because it has not been measured prior to the solution of the problem. There are three possibilities: (i) wait for the conductivity to be measured with the necessary accuracy (a conservative solution, but impractical in many engineering situations); (ii) solve the equation for many different values of the conductivity to get an overall idea of the behavior of the solution (also impractical when the number of parameters increases); or (iii) solve the heat equation only once for any value of the conductivity in a given range, thus providing a sort of abacus or general solution. Obviously the third alternative is the most exciting one, which we will follow with the development of the PGD. To compute this general solution, however, it suffices to introduce the conductivity as an extra-coordinate of the problem, playing the same role as the standard space and time coordinates. The PGD separated approximation of the general solution would then be of the form

$$u(\mathbf{x}, t, k) = \sum_{i=1}^{N} X_i(\mathbf{x}) \cdot T_i(t) \cdot K_i(k), \tag{1.21}$$

wherein the unknown functions $K_i(k)$ are defined in the domain Ω_k of values for the thermal conductivity.

This procedure works well in practice, and it can be extended to introduce many other extra-coordinates, such as source terms, boundary conditions, initial conditions, and geometrical parameters defining the problem's spatial domain. The price to pay is the corresponding increase of the dimensionality of the resulting model that now contains the standard physical coordinates (space and time) plus all the other extra-coordinates that we decided to introduce. We have seen, however, that the PGD handles without difficulty this increased dimensionality. A catalogue of demonstrators can be found in [24–26] and the references therein. With this approach, wherein parameters are considered advantageously as new coordinates of the model, *a posteriori* inverse identification or optimization can be carried out easily as illustrated later in this book. Moreover, if the parameters have a stochastic nature with a given probability distribution, a probability distribution or the desired moments of the model solution can be obtained from the parameter distribution.

This new PGD framework enables us to perform such calculations very efficiently, since the envelope containing all possible solutions has been computed offline and only once, in the form of a parametric solution expressed in a separated form (e.g. the general solution $u(\mathbf{x}, t, k)$ of the above example) that circumvents the curse of dimensionality. Thus, further calculations performed in subsequent optimization or inverse analyses only involve the *post-processing* of the pre-calculated parametric solution. This fact opens almost unimaginable possibilities in simulation-based real-time control: using the parametric solution computed offline as a suitable transfer function, one could proceed to online control and augmented reality with extremely light computing platforms like tablets or smartphones. Consider for example the case of haptic devices discussed previously. Their high-frequency operational requirements (more than 500 Hz) would thus be met by the real-time post-processing of the PGD solution of the problem, with the applied load and its point of application being treated as extra-coordinates.

As can be inferred from these introductory comments, the Proper Generalized Decomposition has been successfully applied to a plethora of different problems ranging from the original space-time representations to more recent applications wherein models are efficiently cast into a high-dimensional framework to allow for efficient real-time post-processing of the PGD solution. The potential field of applications of the PGD does not seem, however, to have been fully explored. We believe indeed that there is a lot of room in the PGD field to perform further research, both at the theoretical level and from the point of view of applications.

1.5 A Brief Overview of PGD Applications

1.5.1 Multidimensional Models

As discussed previously, many models in science and engineering are inherently defined in a mathematical space that is by nature high-dimensional. Their numerical discretization leads to the well-known curse of dimensionality if traditional, mesh-based techniques are applied. Separated representations, on the contrary, provide us with an efficient means of dealing with these requirements.

Separated representations were successfully applied for solving the multidimensional Fokker-Planck equation describing complex fluids in the framework of kinetic theory. In [27], we addressed the solution of the linear Fokker-Planck equation describing multi-bead-spring molecular models of polymer solutions in steady-state homogeneous flows. In this context, the unknown field is the distribution function $\Psi(\mathbf{q}_1, \ldots, \mathbf{q}_D)$ describing the molecular conformation $(\mathbf{q}_1, \ldots, \mathbf{q}_D)$ of individual polymers, as illustrated in Fig. 1.5. The distribution function evolves according to the associated Fokker-Planck equation

$$\frac{\partial \Psi}{\partial t} = -\sum_{i=1}^{D} (\dot{\mathbf{q}}_i \cdot \Psi) . \tag{1.22}$$

Here, the conformational velocities $\dot{\mathbf{q}}_i$ depend on conformation and polymer-solvent interactions, so that the Fokker-Planck equation takes the final form of a convection-diffusion equation in conformation space. The solution was sought in the separated form

$$\Psi(\mathbf{q}_1, \ldots, \mathbf{q}_D) = \sum_{i=1}^{N} F_i^1(\mathbf{q}_1) \times \cdots \times F_i^D(\mathbf{q}_D). \tag{1.23}$$

The above PGD solution procedure was extended to non-linear kinetic theory descriptions of complex fluids in [28]. The transient case was addressed in [29]. A deeper analysis of non-linear and transient models was considered in [30]. Extension of these developments to the simulation of complex flows was performed in [31–33], thus opening very encouraging perspectives and claiming the necessity of defining

Fig. 1.5 Molecular conformation

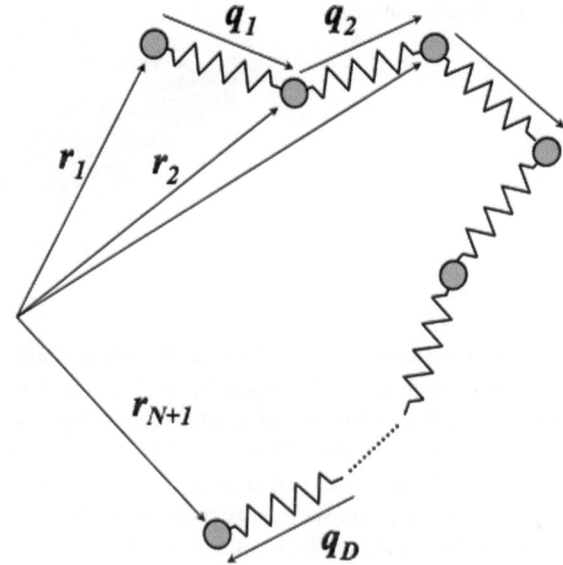

efficient stabilization techniques. Finally, the PGD was implemented in [34] to solve the stochastic equation within the framework of Brownian Configuration Fields. Models involving suspensions and colloidal systems were considered in [35–40] and kinetic descriptions of microstructural mixing in [41, 42].

The interested reader can consult [43] for an exhaustive overview of PGD applications in computational rheology.

Multidimensional models encountered in the finer descriptions of matter (ranging from quantum chemistry to statistical mechanics) were revisited in [41]. The multidimensional chemical master equation was efficiently solved in [7] and [44]. The Langer's equation governing phase transitions was considered in [45].

1.5.2 Separating the Physical Space

Many useful mathematical models are not inherently defined in a high-dimensional space, but they can nevertheless be treated efficiently in a separated manner. Models defined in cubic domains, or hypercubic domains of moderate dimension, suggest the following separated representation

$$u(x, y, z) = \sum_{i=1}^{N} X_i(x) \cdot Y_i(y) \cdot Z_i(z). \tag{1.24}$$

Fig. 1.6 Honeycomb plate

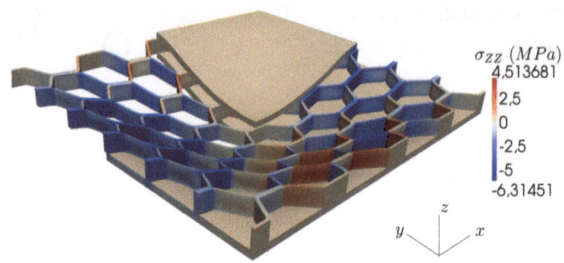

Thus, the PGD computation of a 3D solution involves a sequence of 1D solutions for computing the functions $X_i(x)$, $Y_i(y)$ and $Z_i(z)$. This is highly relevant in homogenization problems wherein the elementary volume element is a cube. The interested reader can refer to [46] or [47]. Fully-separated representations in complex, non-hypercubic domains can also be performed [48].

In the case of plate, shell or extruded geometries, one could also consider advantageously the separated approximation [49, 50]:

$$u(x, y, z) = \sum_{i=1}^{N} X_i(x, y) \cdot Z_i(z). \tag{1.25}$$

For plate geometries, (x, y) are the in-plane coordinates and z is the thickness coordinate. In the case of extruded profiles, (x, y) represents the surface extruded in the z direction. This PGD representation makes it possible to compute fully-3D solutions at a numerical cost characteristic of 2D solutions, without any simplifying a priori assumption. Figure 1.6 illustrates the analysis of a complex plate structure by using the separated representation just described. Interestingly, as discussed in [49], the first mode of the PGD decomposition is very close to the standard plate and shell solutions, and additional modes refine such a solution, thus taking into account 3D effects. The first steps towards addressing shell geometries with the PGD are reported in [51] and a fine analysis of cracked plates in [52].

1.5.3 Parametric Models Defining Virtual Charts: A Route to Efficient Optimization, Inverse Identification and Real-time Simulation

High-dimensional models such as those arising from kinetic theory of complex fluids may seem far from the usual practice of computational science. As discussed above with the example of a parametric thermal model, however, more standard models of low or moderate dimensionality can easily be enriched by introducing several of the problem parameters as extra-coordinates. In so doing, the PGD strategy can be

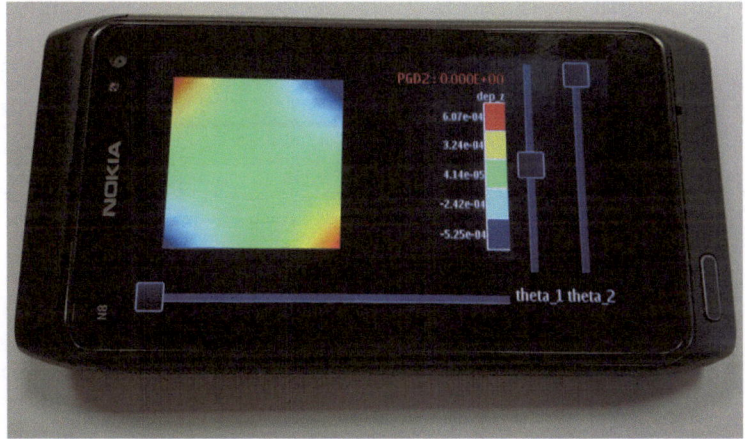

Fig. 1.7 Composite laminate analysis on a smartphone

harnessed to produce a general solution of great value for further processing such as optimization or inverse identification.

This new approach to parametric modelling was addressed in [49, 53, 54] where material parameters were introduced as extra-coordinates. In [55] thermal conductivities, macroscopic temperature and its time evolution were introduced as extra-coordinates for performing linear and non-linear homogenization. In [49], the anisotropy directions of plies involved in a composite laminate were considered as extra-coordinates. By assuming a given level of uncertainty in the actual fiber orientation, the authors evaluated the envelope of the resulting distorted structures due to the thermomechanical coupling.

Moreover, once the PGD separated representation of the parametric solution has been computed offline, its online use only requires one to particularize the parametric solution for a desired set of parameter values. Obviously, this task can be performed very fast and repeatedly in real-time, by using light computing platforms such as smartphones or tablets. Figure 1.7 illustrates one of those applications [49] in which the elastic solution of a two-ply composite laminate was computed by introducing the fiber orientation θ_1 and θ_2 in each ply as extra-coordinates:

$$u_j(x, y, z, \theta_1, \theta_2) = \sum_{i=1}^{N} X_i^j(x, y) \cdot Z_i^j(z) \cdot \Theta_i^{j,1}(\theta_1) \cdot \Theta_i^{j,2}(\theta_2). \tag{1.26}$$

One can then visualize each component u_j of the displacement field by particularizing the z-coordinate as well as the orientation of the reinforcement in both plies.

Process optimization was considered in [56, 57], for instance. Shape optimization was performed in [58] by considering the layer thicknesses of a composite laminate as extra-coordinates, leading to the model solution in any of the geometries generated by the parameters considered as extra-coordinates.

Fig. 1.8 Parameterized domain

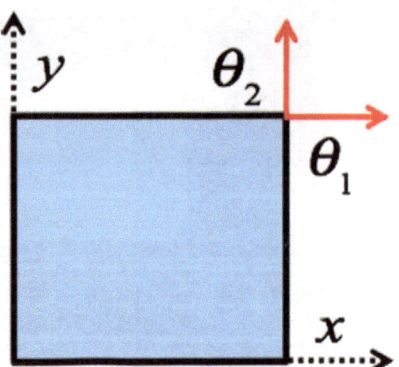

More complex scenarios were considered in [59], where the Laplace equation was solved in the parameterized domain depicted in Fig. 1.8. The PGD general solution was sought in the form

$$u(x, y, \theta_1, \theta_2) = \sum_{i=1}^{N} X_i(x, y) \cdot \Theta_i^1(\theta) \cdot \Theta_i^2(\theta_2). \qquad (1.27)$$

Then, the general solution was particularized for two particular choices of domain geometry, i.e. of θ_1 and θ_2 as illustrated in Fig. 1.9.

1.5.4 Real-time Simulation, DDDAS and More

It is easy to understand that after performing PGD computations wherein problem parameters are considered as extra-coordinates of the model, *a posteriori* inverse

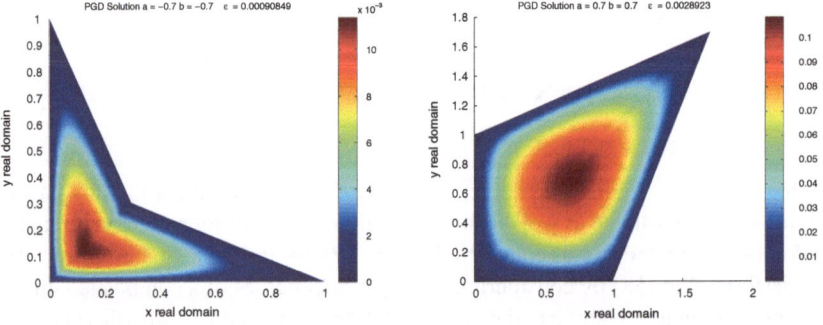

Fig. 1.9 Solution in two different geometries

identification or optimization can easily be handled. The PGD framework indeed allows one to perform these tasks very efficiently as simple post-processing operations acting on the PGD general solution computed once and for all.

Inverse methods were addressed in [60] in the context of real-time simulations. They were coupled with control strategies in [61] as a first step towards DDDAS. Moreover, because the PGD parametric solution is pre-computed offline, it can be used online in real-time mode using light computing platforms like smartphones [49, 61]. This constitutes a first step towards the use of this kind of information in augmented reality platforms.

As mentioned before, surgical simulators must operate at frequencies higher than 500 Hz. The use of model reduction seems to be an appealing alternative for reaching such performances. POD-based techniques, however, even when combined with an asymptotic numerical methods to avoid the computation of the tangent matrix, exhibit serious difficulties to fulfill such requirements [62, 63]. On the other hand, the PGD parametric solution for the three components $u_j(\mathbf{x}, \mathbf{p}, \mathbf{y})$ of the displacement vector, wherein the applied load $\mathbf{p}$ and its point of application $\mathbf{y}$ are considered as extra-coordinates, can be computed offline in the form

$$u_j(\mathbf{x}, \mathbf{p}, \mathbf{y}) = \sum_{i=1}^{N} X_i^j(\mathbf{x}) \cdot P_i^j(\mathbf{p}) \cdot Y_i^j(\mathbf{y}), \tag{1.28}$$

and then later particularized online as explained above. This approach opens a wide field of applications [64, 65]. An illustration is given in Fig. 1.10.

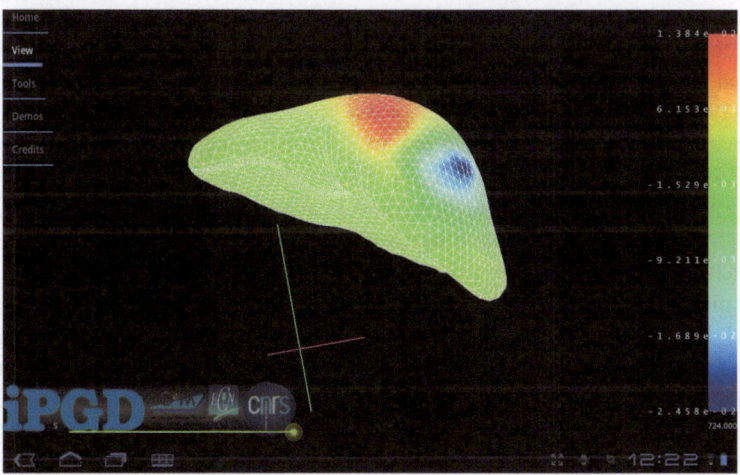

Fig. 1.10 Towards real-time surgical simulations

1.5.5 Open Issues and Further Developments

There is a variety of open questions related to PGD-based discretization techniques.

Stabilization is one of the main issues, and concerns both convective terms and mixed formulations. It is well known that advection-dominated equations require appropriate numerical stabilization (upwinding) in order to avoid the spurious oscillations induced by centered discretization schemes. Standard upwinding stabilization schemes must be adapted to separated representations and multidimensional advection-diffusion equations [48]. On the other hand the treatment of mixed formulations involving many fields whose approximations are constrained by the so-called LBB stability conditions must be addressed carefully [31, 66].

Another important question is that of error estimation. First attempts were considered in [67–69]. Coupling standard mesh-based discretization techniques with reduced bases (POD and PGD) is an issue of major interest in order to represent localized behaviors (e.g. discontinuities, singularities, and boundary layers). Some coupling strategies were proposed in [70] and [71].

Multi-scale and multi-physics, non-linearly coupled models involving different characteristic times were efficiently coupled in [72]; a globalization of local problems to speed-up the simulation in a separated representation framework was proposed. In [73], the time axis was transformed into a two-dimensional domain to account for two very different time scales. Parallel time integration strategies, inspired from the ones proposed in [46], were considered in [74] within the PGD framework. The coupling of other multi-physics models in the context of composites manufacturing processes was performed in [56, 75]. Non-incremental parametric solutions defined in evolving domains were addressed in [76]. In [77], the solution of the Maxwell equations of electromagnetism was considered. Uncertainty quantification and propagation have been deeply considered in the works of A. Nouy. The interested reader can refer to [78] and the references therein.

In [79], the PGD was introduced in the boundary element method framework for solving transient models where few works concerning the use of reduced bases exist [17]. In view of the space-time separated representation, only the steady-state kernel (and not the time-dependent kernel) is needed in the PGD formalism. This yields significant computational time savings. Finally, we considered in [80] a PGD approach to augmented learning in science and engineering higher education.

Despite the numerous recent works concerning the foundations and applications of the PGD in many branches of engineering sciences, this technique remains young and it faces many challenging scenarios that require further developments. Among those, we could cite the optimality and robustness of separated representation constructors, a topic that requires further research in numerical analysis; domain partitioning (some preliminary but promising results exist by considering the Arlequin technique [81] within the PGD framework); solid mechanics with large material transformations (in that case, space and time cannot be directly separated); hyperbolic problems; non-linear multi parametric models; how to address discontinuities and localization; how to define bridges between the PDG and standard commercial simulation codes.

Even if this list is not exhaustive, these are some hot topics whose solution could open new routes towards advanced simulation in engineering sciences.

References

1. E. Cancès, M. Defranceschi, W. Kutzelnigg, C. Le Bris, Y. Maday, Computational Quantum Chemistry: a primer. in *Handbook of Numerical Analysis*, vol. 10 (Elsevier, Amsterdam, 2003) pp. 3–270
2. C. Le Bris (ed) Handbook of Numerical Analysis, in *Computational Chemistry*, vol. 10 (Elsevier, Amsterdam, 2003)
3. C. Le Bris, P. L. Lions, From atoms to crystals: a mathematical journey. Bull. Am. Math. Soc. **42**(3), 291–363 (2005)
4. C. Le Bris, Mathematical and numerical analysis for molecular simulation: accomplishments and challenges, in *Proceedings of the International Congress of Mathematicians*, Madrid, Spain, 2003, pp. 1507–1522
5. B.B. Bird, C.F. Curtiss, R.C. Armstrong, O. Hassager, Dynamics of polymeric liquids, in *Kinetic Theory*, vol. 2 (Wiley, New York, 1987)
6. N. Bellomo, *Modeling Complex Living Systems* (Birkhauser, Boston, 2008)
7. A. Ammar, E. Cueto, F. Chinesta, Reduction of the chemical master equation for gene regulatory networks using proper generalized dcompositions. Int. J. Numer. Methods Biomed. Eng. **28**(9), 960–973 (2012)
8. F. Darema, Engineering/scientific and commercial applications: differences, similarities, and future evolution. in *Proceedings of the Second Hellenic European Conference on Mathematics and Informatics, HERMIS*, vol. 1, 1994, pp. 367–374
9. N.S.F. Final, D.D.D.A.S. Report, Workshop, (Arlington, VA, 2006)
10. J.T. Oden, T. Belytschko, J. Fish, T.J.R. Hughes, C. Johnson, D. Keyes, A. Laub, L. Petzold, D. Srolovitz, S.Yip. Simulation-Based Engineering Science: Revolutionizing Engineering Science through simulation. NSF Blue Ribbon Panel on SBES, 2006
11. D. Ryckelynck, F. Chinesta, E. Cueto, A. Ammar, On the a priori model reduction: overview and recent developments. Arch. Comput. methods Eng. State Art Rev. **13**(1), 91–128 (2006)
12. R.A. Bialecki, A.J. Kassab, A. Fic, Proper orthogonal decomposition and modal analysis for acceleration of transient FEM thermal analysis. Int. J. Numer. Meth. Eng. **62**, 774–797 (2005)
13. J. Burkardt, M. Gunzburger, H-Ch. Lee, POD and CVT-based reduced-order modeling of Navier-Stokes flows. Comput. Methods Appl. Mech. Eng. **196**, 337–355 (2006)
14. M.D. Gunzburger, J.S. Peterson, J.N. Shadid, Reduced-order modeling of time-dependent PDEs with multiple parameters in the boundary data. Comput. Methods Appl. Mech. Eng. **196**, 1030–1047 (2007)
15. Y. Maday, E.M. Ronquist, The reduced basis element method: application to a thermal fin problem. SIAM J. Sci. Comput., **26**(1), 240–258 (2004)
16. H.M. Park, D.H. Cho, The use of the Karhunen-Loeve decomposition for the modelling of distributed parameter systems. Chem. Eng. Sci. **51**, 81–98 (1996)
17. D. Ryckelynck, L. Hermanns, F. Chinesta, E. Alarcon, An efficient a priori model reduction for boundary element models. Eng. Anal. Bound. Elem. **29**, 796–801 (2005)
18. D. Ryckelynck, F. Chinesta, E. Cueto, A. Ammar, On the a priori model reduction: overview and recent developments. Arch. Comput. Methods Eng. **12**(1), 91–128 (2006)
19. P. Ladevèze, The large time increment method for the analyze of structures with nonlinear constitutive relation described by internal variables. C. R. Acad. Sci. Paris **309**, 1095–1099 (1989)
20. P. Ladevèze, J.-C. Passieux, D. Néron, The latin multiscale computational method and the proper generalized decomposition. Comput. Methods Appl. Mech. Eng. **199**(21–22), 1287–1296 (2010)

21. D. Néron, P. Ladevèze, Proper generalized decomposition for multiscale and multiphysics problems. Arch. Comput. Methods Eng. **17**(4), 351–372 (2010)
22. A. Nouy, P. Ladevèze, Multiscale computational strategy with time and space homogenization: a radial-type approximation technique for solving microproblems. Int. J. Multiscale Comput. Eng. **170**(2), 557–574 (2004)
23. J.-C. Passieux, P. Ladevèze, D.Néron, A scalable time-space multiscale domain decomposition method: adaptive time scale separation. Comput. Mech. **46**(4), 621–633 (2010)
24. F. Chinesta, A. Ammar, E. Cueto, Recent advances and new challenges in the use of the proper generalized decomposition for solving multidimensional models. Arch. Comput. Methods Eng. **17**(4), 327–350 (2010)
25. F. Chinesta, P. Ladeveze, E. Cueto, A short review in model order reduction based on proper generalized decomposition. Arch. Comput. Methods Eng. **18**, 395–404 (2011)
26. F. Chinesta, A. Leygue, F. Bordeu, J.V. Aguado, E. Cueto, D. Gonzalez, I. Alfaro, A. Ammar, A. Huerta, PGD-based computational vademecum for efficient design, optimization and control. Arch. Comput. Methods Eng. **20**, 31–59 (2013)
27. A. Ammar, B. Mokdad, F. Chinesta, R. Keunings, A new family of solvers for some classes of multidimensional partial differential equations encountered in kinetic theory modeling of complex fluids. J. Nonnewton. Fluid Mech. **139**, 153–176 (2006)
28. B. Mokdad, E. Pruliere, A. Ammar, F. Chinesta, On the simulation of kinetic theory models of complex fluids using the Fokker-Planck approach. Appl. Rheol. **17**(2), 1–14 (2007) 26494
29. A. Ammar, B. Mokdad, F. Chinesta, R. Keunings, A new family of solvers for some classes of multidimensional partial differential equations encountered in kinetic theory modeling of complex fluids. Part II: Transient simulation using space-time separated representation. J. Nonnewton. Fluid Mech. **144**, 98–121 (2007)
30. A. Ammar, M. Normandin, F. Daim, D. Gonzalez, E. Cueto, F. Chinesta, Non-incremental strategies based on separated representations: applications in computational rheology. Commun. Math. Sci. **8**(3), 671–695 (2010)
31. S. Aghighi, A. Ammar, C. Metivier, M. Normandin, F. Chinesta, Non-incremental transient solution of the Rayleigh-Bnard convection model using the PGD. J. Nonnewton. Fluid Mech. **200**, 65–78 (2013). doi: 10.1016/j.jnnfm.2012.11.007 (2013)
32. B. Mokdad, A. Ammar, M. Normandin, F. Chinesta, J.R. Clermont, A fully deterministic micro-macro simulation of complex flows involving reversible network fluid models. Math. Comput. Simul. **80**, 1936–1961 (2010)
33. E. Pruliere, A. Ammar, N. El Kissi, F. Chinesta, Recirculating flows involving short fiber suspensions: numerical difficulties and efficient advanced micro-macro solvers. Arch. Comput. Methods Eng State Art Rev. **16**, 1–30 (2009)
34. F. Chinesta, A. Ammar, A. Falco, M. Laso, On the reduction of stochastic kinetic theory models of complex fluids. Model. Simul. Mater. Sci. Eng. **15**, 639–652 (2007)
35. E. Abisset-Chavanne, R. Mezher, S. Le Corre, A. Ammar, F. Chinesta. Kinetic theory microstructure modeling in concentrated suspensions. Entropy, In press
36. F. Chinesta, From single-scale to two-scales kinetic theory descriptions of rods suspensions. Arch. Comput. Methods Eng. **20**(1), 1–29 (2013)
37. M. Grmela, F. Chinesta, A. Ammar, Mesoscopic tube model of fluids composed of wormlike micelles. Rheol. Acta **49**(5), 495–506 (2010)
38. M. Grmela, A. Ammar, F. Chinesta. One and two-fiber orientation kinetic theories of fiber suspensions. J. Nonnewton. Fluid Mech. **200**, 17–33 (2013). doi: 10.1016/j.jnnfm.2012.10.009 (2013)
39. G. Maitejean, A. Ammar, F Chinesta, M. Grmela, Deterministic solution of the kinetic theory model of colloidal suspensions of structureless particles. Rheol. Acta **51**(6), 527–543 (2012)
40. G. Maitrejean, M. Grmela, A. Ammar, F. Chinesta, Kinetic theory of colloidal suspensions: morphology, rheology and migration. Rheol. Acta **52**(6), 557–577 (2013)
41. A. Ammar, F. Chinesta, P. Joyot, The nanometric and micrometric scales of the structure and mechanics of materials revisited: an introduction to the challenges of fully deterministic numerical descriptions. Int. J. Multiscale Comput. Eng. **6**(3), 191–213 (2008)

42. G. Maitrejean, A. Ammar, F. Chinesta, Simulating microstructural evolution during pasive mixing. Int. J. Mater. Form. **5**(1), 73–81 (2012)
43. F. Chinesta, A. Ammar, A. Leygue, R. Keunings, An overview of the proper generalized decomposition with applications in computational rheology. J. Nonnewton. Fluid Mech. **166**, 578–592 (2011)
44. F. Chinesta, A. Ammar, E. Cueto, On the use of proper generalized decompositions for solving the multidimensional chemical master equation. Eur. J. Comput. Mech. **19**, 53–64 (2010)
45. H. Lamari, A. Ammar, A. Leygue, F. Chinesta, On the solution of the multidimensional Langer's equation by using the proper generalized decomposition method for modeling phase transitions. Model. Simul. Mater. Sci. Eng. **20**, 015007 (2012)
46. F. Chinesta, A. Ammar, F. Lemarchand, P. Beauchene, F. Boust, Alleviating mesh constraints: model reduction, parallel time integration and high resolution homogenization. Comput. Methods Appl. Mech. Eng. **197**(5), 400–413 (2008)
47. E. Pruliere, F. Chinesta. A. Ammar, A. Leygue, A. Poitou, On the solution of the heat equation in very thin tapes. Int. J. Thermal Sci. **65**, 148–157 (2013)
48. D. Gonzalez, A. Ammar, F. Chinesta, E. Cueto, Recent advances in the use of separated representations. Int. J. Numerical Methods Eng. **81**(5), 637–659 (2010)
49. B. Bognet, A. Leygue, F. Chinesta, A. Poitou, F. Bordeu, Advanced simulation of models defined in plate geometries: 3D solutions with 2D computational complexity. Comput. Methods Appl. Mech. Eng. **201**, 1–12 (2012)
50. A. Leygue, F. Chinesta, M. Beringhier, T.L. Nguyen, J.C. Grandidier, F. Pasavento, B. Schrefler, Towards a framework for non-linear thermal models in shell domains. Int. J. Numerical Methods Heat Fluid Flow, **23**(1), 55–73 (2013)
51. B. Bognet, A. Leygue, F. Chinesta, On the fully 3D simulation of thermoelastic models defined in plate geometries. Eur. J. Comput. Mech. **21**(1–2), 40–51 (2012)
52. E. Giner, B. Bognet, J.J. Rodenas, A. Leygue, J. Fuenmayor, F. Chinesta, The proper generalized decomposition (PGD) as a numerical procedure to solve 3D cracked plates in linear elastic fracture mechanics. Int. J. Solid Structures **50**(10), 1710–1720 (2013)
53. A. Ammar, M. Normandin, F. Chinesta, Solving parametric complex fluids models in rheometric flows. J. Nonnewton. Fluid Mech. **165**, 1588–1601 (2010)
54. E. Pruliere, F. Chinesta, A. Ammar, On the deterministic solution of multidimensional parametric models by using the proper generalized decomposition. Math. Comput. Simul. **81**, 791–810 (2010)
55. H. Lamari, A. Ammar, P. Cartraud, G. Legrain, F. Jacquemin. F. Chinesta, Routes for efficient computational homogenization of non-linear materials using the proper generalized decomposition. Arch. Comput. Methods Eng. **17**(4), 373–391 (2010)
56. F. Chinesta, A. Leygue, B. Bognet, Ch. Ghnatios, F. Poulhaon, F. Bordeu, A. Barasinski, A. Poitou, S. Chatel, S. Maison-Le-Poec, First steps towards an advanced simulation of composites manufacturing by Automated Tape Placement. Int. J. Mater. Form. doi: 10.1007/s12289-012-1112-9 (in press)
57. Ch. Ghnatios, F. Chinesta, E. Cueto, A. Leygue, P. Breitkopf, P. Villon, Methodological approach to efficient modeling and optimization of thermal processes taking place in a die: application to pultrusion. Compos. Part A **42**, 1169–1178 (2011)
58. A. Leygue, E. Verron, A first step towards the use of proper general decomposition method for structural optimization. Arch. Comput. Methods Eng. **17**(4), 465–472 (2010)
59. A. Ammar, A. Huerta, F. Chinesta, E. Cueto, A. Leygue, Parametric solutions involving geometry: a step towards efficient shape optimization. Comput. Methods Appl. Mech. Eng. doi: 10.1016/j.cma.2013.09.003 (in press)
60. D. Gonzalez, F. Masson, F. Poulhaon, A. Leygue, E. Cueto, F. Chinesta, Proper generalized decomposition based dynamic data-driven inverse identification. Math. Comput. Simul. **82**(9), 1677–1695 (2012)
61. Ch. Ghnatios, F. Masson, A. Huerta, E. Cueto, A. Leygue, F. Chinesta, Proper generalized decomposition based dynamic data-driven control of thermal processes. Comput. Methods Appl. Mech. Eng. **213**, 29–34 (2012)

62. S. Niroomandi, I. Alfaro, E. Cueto, F. Chinesta, Real-time deformable models of non-linear tissues by model reduction techniques. Comput. Meth. Programs Biomed. **91**, 223–231 (2008)
63. S. Niroomandi, I. Alfaro, D. Gonzalez, E. Cueto, F. Chinesta, Real time simulation of surgery by reduced order modelling and X-FEM techniques. Int. J. Numerical Methods Biomed. Eng. **28**(5), 574–588 (2012)
64. S. Niroomandi, D. Gonzalez, I. Alfaro, F. Bordeu, A. Leygue, E. Cueto, F. Chinesta, Real time simulation of biological soft tissues : a PGD approach. Int. J. Numerical Methods Biomed. Eng. **29**(5), 586–600 (2013)
65. I. Alfaro, D. Gonzalez, F. Bordeu, A. Leygue, A. Ammar, E. Cueto, F. Chinesta, Real-time in sillico experiments on gene regulatory networks and surgery simulation on handheld devices, J. Comput. Surg. (in press)
66. A. Dumon, C. Allery, A. Ammar, Proper generalized decomposition method for incompressible Navier-Stokes equations with a spectral discretization. Appl. Math. Comput. **219**(15), 8145–8162 (2013)
67. A. Ammar, F. Chinesta, P. Diez, A. Huerta, An error estimator for separated representations of highly multidimensional models. Comput. Methods Appl. Mech. Eng. **199**, 1872–1880 (2010)
68. P. Ladevèze, L. Chamoin, On the verification of model reduction methods based on the proper generalized decomposition. Comput. Methods Appl. Mech. Eng. **200**, 2032–2047 (2011)
69. J.P. Moitinho, A basis for bounding the errors of proper generalised decomposition solutions, in solid mechanics. Int. J. Numerical Methods Eng. **94**(10), 961–981 (2013)
70. A. Ammar, E. Pruliere, J. Ferec, F. Chinesta, E. Cueto, Coupling finite elements and reduced approximation bases. Eur. J. Comput. Mech. **18**(5–6), 445–463 (2009)
71. A. Ammar, F. Chinesta, E. Cueto, Coupling finite elements and proper generalized decomposition. Int. J. Multiscale Comput. Eng. **9**(1), 17–33 (2011)
72. F. Chinesta, A. Ammar, E. Cueto, Proper generalized decomposition of multiscale models. Int. J. Numerical Methods Eng. **83**(8–9), 1114–1132 (2010)
73. A. Ammar, F. Chinesta, E. Cueto, M. Doblare, Proper generalized decomposition of time-multiscale models. Int. J. Numerical Methods Eng. **90**(5), 569–596 (2012)
74. F. Poulhaon, F. Chinesta, A. Leygue, A first step towards a PGD based parallelization strategy. Eur. J. Comput. Mech. **21**(3–6), 300–311 (2012)
75. E. Pruliere, J. Ferec, F. Chinesta, A. Ammar, An efficient reduced simulation of residual stresses in composites forming processes. Int. J. Mater. Form. **3**(2), 1339–1350 (2010)
76. A. Ammar, E. Cueto, F. Chinesta, Non-incremental PGD solution of parametric uncoupled models defined in evolving domains. Int. J. Numerical Methods Eng. **93**(8), 887–904 (2013)
77. M. Pineda, F. Chinesta, J. Roger, M. Riera, J. Perez, F. Daim, Simulation of skin effect via separated representations. Int. J. Comput. Math. Electr. Electron. Eng. **29**(4), 919–929 (2010)
78. A. Nouy, Proper generalized decompositions and separated representations for the numerical solution of high dimensional stochastic problems. Arch. Comput.l Methods Eng. State Art Rev. **17**, 403–434 (2010)
79. G. Bonithon, P. Joyot, F. Chinesta, P. Villon, Non-incremental boundary element discretization of parabolic models based on the use of proper generalized decompositions. Eng. Anal. Bound. Elem. **35**(1), 2–17 (2011)
80. F. Bordeu, A. Leygue, D. Modesto, D. Gonzalez, E. Cueto, F. Chinesta, A PGD submitted to augmented learning and science and engineering high education. J. Eng. Educ
81. H. Ben Dhia, Multiscale mechanical problems: the Arlequin method. C. R. Acad. Sci. Paris Ser-II b **326**, 899–904 (1998)

Chapter 2
PGD Solution of the Poisson Equation

Abstract This chapter describes the main features of the PGD technique, in particular the one related to the construction of a separated representation of the unknown
field involved in a partial differential equation. For this purpose, we consider the
solution of the two-dimensional Poisson equation in a square domain. The solution is sought as a finite sum of terms, each one involving the product of functions
of each coordinate. The solution is then calculated by means of a sequence of one-
dimensional problems. The chapter starts with the simplest case, that is later extended
to cover more complex problems: non-constant source terms, non-homogeneous
Dirichlet and Neumann boundary conditions, and high-dimensional problems. Carefully solved numerical examples are discussed to illustrate the theoretical developments.

Keywords Multidimensional model · Poisson's problem · Proper Generalized
Decomposition

It is now time to detail the inner workings of the PGD. We begin with a simple but
illustrative case study, which we shall progressively make more complex.

Consider the solution of the Poisson equation

$$\Delta u(x, y) = f(x, y), \tag{2.1}$$

in a two-dimensional rectangular domain $\Omega = \Omega_x \times \Omega_y = (0, L) \times (0, H)$.

We specify homogeneous Dirichlet boundary conditions for the unknown field
$u(x, y)$, i.e. $u(x, y)$ vanishes at the domain boundary Γ. Furthermore, we assume
that the source term f is constant over the domain Ω.

For all suitable test functions u^*, the weighted residual form of (2.1) reads

$$\int_{\Omega_x \times \Omega_y} u^* \cdot (\Delta u - f) \, dx \cdot dy = 0, \tag{2.2}$$

F. Chinesta et al., *The Proper Generalized Decomposition for Advanced Numerical*   25
Simulations, SpringerBriefs in Applied Sciences and Technology,
DOI: 10.1007/978-3-319-02865-1_2, © The Author(s) 2014

or more explicitly

$$\int_{\Omega_x \times \Omega_y} u^* \cdot \left(\frac{\partial^2 u}{\partial x^2} + \frac{\partial^2 u}{\partial y^2} - f \right) \, dx \cdot dy = 0. \tag{2.3}$$

Our goal is to obtain a PGD approximate solution to (2.1) in the separated form

$$u(x, y) = \sum_{i=1}^{N} X_i(x) \cdot Y_i(y). \tag{2.4}$$

We shall do so by computing each term of the expansion one at a time, thus enriching the PGD approximation until a suitable convergence criterion is satisfied.

2.1 Progressive Construction of the Separated Representation

At each enrichment step n ($n \geq 1$), we have already computed the $n - 1$ first terms of the PGD approximation (2.4):

$$u^{n-1}(x, y) = \sum_{i=1}^{n-1} X_i(x) \cdot Y_i(y). \tag{2.5}$$

We now wish to compute the next term $X_n(x) \cdot Y_n(y)$ to obtain the enriched PGD solution

$$u^n(x, y) = u^{n-1}(x, y) + X_n(x) \cdot Y_n(y) = \sum_{i=1}^{n-1} X_i(x) \cdot Y_i(y) + X_n(x) \cdot Y_n(y). \tag{2.6}$$

Both functions $X_n(x)$ and $Y_n(y)$ are unknown at the current enrichment step n, and they appear in the form of a product. The resulting problem is thus non-linear and a suitable iterative scheme is required. We shall use the index p to denote a particular iteration.

At enrichment step n, the PGD approximation $u^{n,p}$ obtained at iteration p thus reads

$$u^{n,p}(x, y) = u^{n-1}(x, y) + X_n^p(x) \cdot Y_n^p(y). \tag{2.7}$$

The simplest iterative scheme is an alternating direction strategy that computes $X_n^p(x)$ from $Y_n^{p-1}(y)$, and then $Y_n^p(x)$ from $X_n^p(x)$. An arbitrary initial guess $Y_n^0(y)$ is specified to start the iterative process. The non-linear iterations proceed until reaching a fixed point within a user-specified tolerance ϵ, i.e.

$$\frac{\|X_n^p(x) \cdot Y_n^p(y) - X_n^{p-1}(x) \cdot Y_n^{p-1}(y)\|}{\|X_n^{p-1}(x) \cdot Y_n^{p-1}(y)\|} < \epsilon, \tag{2.8}$$

where $\| \cdot \|$ is a suitable norm.

The enrichment step n thus ends with the assignments $X_n(x) \leftarrow X_n^p(x)$ and $Y_n(y) \leftarrow Y_n^p(y)$.

The enrichment process itself stops when an appropriate measure of error $\mathcal{E}(n)$ becomes small enough, i.e $\mathcal{E}(n) < \tilde{\epsilon}$. Several stopping criteria are available, as we shall discuss later. We now describe in more detail one particular alternating direction iteration at a given enrichment step.

2.1.1 Alternating Direction Strategy

Each iteration of the alternating direction scheme consists in the following two steps:

- Calculating $X_n^p(x)$ from $Y_n^{p-1}(y)$

  In this case, the approximation reads

  $$u^{n,p}(x, y) = \sum_{i=1}^{n-1} X_i(x) \cdot Y_i(y) + X_n^p(x) \cdot Y_n^{p-1}(y), \tag{2.9}$$

  where all functions are known except $X_n^p(x)$.

  The simplest choice for the weight function u^* in the weighted residual formulation (2.3) is

  $$u^*(x, y) = X_n^*(x) \cdot Y_n^{p-1}(y), \tag{2.10}$$

  which amounts to select the Galerkin weighted residual form of the Poisson equation.

  Injecting (2.9) and (2.10) into (2.3), we obtain

  $$\int_{\Omega_x \times \Omega_y} X_n^* \cdot Y_n^{p-1} \cdot \left(\frac{d^2 X_n^p}{dx^2} \cdot Y_n^{p-1} + X_n^p \cdot \frac{d^2 Y_n^{p-1}}{dy^2} \right) dx \cdot dy$$

  $$= - \int_{\Omega_x \times \Omega_y} X_n^* \cdot Y_n^{p-1} \cdot \sum_{i=1}^{n-1} \left(\frac{d^2 X_i}{dx^2} \cdot Y_i + X_i \cdot \frac{d^2 Y_i}{dy^2} \right) dx \cdot dy$$

  $$+ \int_{\Omega_x \times \Omega_y} X_n^* \cdot Y_n^{p-1} \cdot f \, dx \cdot dy. \tag{2.11}$$

Here comes a crucial point: since all functions of y are known in the above expression, we can compute the following one-dimensional integrals over Ω_y:

$$
\begin{cases}
\alpha^x = \int_{\Omega_y} \left(Y_n^{p-1}(y) \right)^2 dy \\[2mm]
\beta^x = \int_{\Omega_y} Y_n^{p-1}(y) \cdot \dfrac{d^2 Y_n^{p-1}(y)}{dy^2} \, dy \\[2mm]
\gamma_i^x = \int_{\Omega_y} Y_n^{p-1}(y) \cdot Y_i(y) \, dy \\[2mm]
\delta_i^x = \int_{\Omega_y} Y_n^{p-1}(y) \cdot \dfrac{d^2 Y_i(y)}{dy^2} \, dy \\[2mm]
\xi^x = \int_{\Omega_y} Y_n^{p-1}(y) \cdot f \, dy
\end{cases}
\tag{2.12}
$$

Equation (2.11) becomes

$$
\int_{\Omega_x} X_n^* \cdot \left(\alpha^x \cdot \frac{d^2 X_n^p}{dx^2} + \beta^x \cdot X_n^p \right) dx
$$

$$
= - \int_{\Omega_x} X_n^* \cdot \sum_{i=1}^{n-1} \left(\gamma_i^x \cdot \frac{d^2 X_i}{dx^2} + \delta_i^x \cdot X_i \right) dx + \int_{\Omega_x} X_n^* \cdot \xi^x \, dx. \tag{2.13}
$$

We have thus obtained the weighted residual form of a one-dimensional problem defined over Ω_x that can be solved (e.g. by the finite element method) to obtain the function X_n^p we are looking for. Alternatively, we can return to the corresponding strong formulation

$$
\alpha^x \cdot \frac{d^2 X_n^p}{dx^2} + \beta^x \cdot X_n^p = - \sum_{i=1}^{n-1} \left(\gamma_i^x \cdot \frac{d^2 X_i}{dx^2} + \delta_i^x \cdot X_i \right) + \xi^x, \tag{2.14}
$$

and solve it numerically by means of any suitable numerical method (e.g. finite differences, pseudo-spectral techniques, ...). The strong form (2.14) is a second-order ordinary differential equation for X_n^p. This is due to the fact that the original Poisson equation involves a second-order x-derivative of the unknown field u.

With either the weighted residual or strong formulations, the homogeneous Dirichlet boundary conditions $X_n^p(x = 0) = X_n^p(x = L) = 0$ are readily specified.

Having thus computed $X_n^p(x)$, we are now ready to proceed with the second step of iteration p.

- Calculating $Y_n^p(y)$ from the just-computed $X_n^p(x)$

The procedure exactly mirrors what we have done above. Indeed, we simply exchange the roles played by all relevant functions of x and y.

The current PGD approximation reads

$$u^{n,P}(x,y) = \sum_{i=1}^{n-1} X_i(x) \cdot Y_i(y) + X_n^P(x) \cdot Y_n^P(y),$$ (2.15)

where all functions are known except $Y_n^P(y)$.

The Galerkin formulation of (2.3) is obtained with the particular choice

$$u^*(x,y) = X_n^P(x) \cdot Y_n^*(y).$$ (2.16)

Then, by introducing (2.15) and (2.16) into (2.3), we get

$$\int_{\Omega_x \times \Omega_y} X_n^P \cdot Y_n^* \cdot \left(\frac{d^2 X_n^P}{dx^2} \cdot Y_n^P + X_n^P \cdot \frac{d^2 Y_n^P}{dy^2} \right) dx \cdot dy$$

$$= -\int_{\Omega_x \times \Omega_y} X_n^P \cdot Y_n^* \cdot \sum_{i=1}^{n-1} \left(\frac{d^2 X_i}{dx^2} \cdot Y_i + X_i \cdot \frac{d^2 Y_i}{dy^2} \right) dx \cdot dy$$

$$+ \int_{\Omega_x \times \Omega_y} X_n^P \cdot Y_n^* \cdot f \, dx \cdot dy.$$ (2.17)

As all functions of x are known, the integrals over Ω_x can be computed to obtain

$$\begin{cases} \alpha^y = \int_{\Omega_x} \left(X_n^P(x) \right)^2 dx \\ \beta^y = \int_{\Omega_x} X_n^P(x) \cdot \frac{d^2 X_n^P(x)}{dx^2} dx \\ \gamma_i^y = \int_{\Omega_x} X_n^P(x) \cdot X_i(x) \, dx \\ \delta_i^y = \int_{\Omega_x} X_n^P(x) \cdot \frac{d^2 X_i(x)}{dx^2} dx \\ \xi^y = \int_{\Omega_x} X_n^P(x) \cdot f \, dx \end{cases}$$ (2.18)

Equation (2.17) becomes

$$\int_{\Omega_y} Y_n^* \cdot \left(\alpha^y \cdot \frac{d^2 Y_n^P}{dy^2} + \beta^y \cdot Y_n^P \right) dy$$

$$= -\int_{\Omega_y} Y_n^* \cdot \sum_{i=1}^{n-1} \left(\gamma_i^y \cdot \frac{d^2 Y_i}{dy^2} + \delta_i^y \cdot Y_i \right) dy + \int_{\Omega_y} Y_n^* \cdot \xi^y \, dy.$$ (2.19)

As before, we have thus obtained the weighted residual form of an elliptic problem defined over Ω_y whose solution is the function $Y_n^P(y)$. Alternatively, the corresponding strong formulation of this one-dimensional problem reads

$$\alpha^y \cdot \frac{d^2 Y_n^p}{dy^2} + \beta^y \cdot Y_n^p = -\sum_{i=1}^{n-1} \left(\gamma_i^y \cdot \frac{d^2 Y_i}{dy^2} + \delta_i^y \cdot Y_i \right) + \xi^y. \qquad (2.20)$$

This again is an ordinary differential equation of the second order, due to the fact that the original Poisson equation involves second-order derivatives of the unknown field with respect to y. With both the weighted residual and strong formulations, the homogeneous Dirichlet boundary conditions $Y_n^p(y = 0) = Y_n^p(y = L) = 0$ are readily specified.

We have thus completed iteration p at enrichment step n.

It is important to realize that the original two-dimensional Poisson equation defined over $\Omega = \Omega_x \times \Omega_y$ has been transformed within the PGD framework into a series of *decoupled one-dimensional problems* formulated in Ω_x and Ω_y.

As we shall detail later, should we consider the Poisson equation defined over a domain of dimension D, i.e. $\Omega_1 \times \Omega_2 \times \cdots \times \Omega_D$, then its PGD solution would similarly involve a series of decoupled one-dimensional problems formulated in each Ω_i. This of course explains why the PGD solution of high-dimensional problems is feasible at all.

2.1.2 Stopping Criterion for the Enrichment Process

The enrichment process itself ends when an appropriate measure of error $\mathcal{E}(n)$ becomes small enough, i.e $\mathcal{E}(n) < \tilde{\epsilon}$. Several stopping criteria are suitable.

A first stopping criterion is associated with the relative weight of the newly-computed term within the PGD expansion. Thus, $\mathcal{E}(n)$ is given by

$$\mathcal{E}(n) = \frac{\|X_n(x) \cdot Y_n(y)\|}{\|u^n(x, y)\|} = \frac{\|X_n(x) \cdot Y_n(y)\|}{\left\| \sum_{i=1}^{n} X_i(x) \cdot Y_i(y) \right\|}. \qquad (2.21)$$

Selecting for example the L^2-norm, we have

$$\|X_n(x) \cdot Y_n(y)\|_2 = \left(\int_{\Omega_x \times \Omega_y} (X_n(x))^2 \cdot (Y_n(y))^2 \, dx \cdot dy \right)^{\frac{1}{2}}$$

$$= \left(\int_{\Omega_x} (X_n(x))^2 \, dx \right)^{\frac{1}{2}} \cdot \left(\int_{\Omega_y} (Y_n(y))^2 \, dy \right)^{\frac{1}{2}}. \qquad (2.22)$$

Since $u^n(x, y)$ is expressed in a separated form, its square will be also expressed in a separated form having $\frac{n \cdot (n+1)}{2}$ terms:

$$(u^n)^2 = \sum_{i=1}^{\frac{n \cdot (n+1)}{2}} S_i^x(x) \cdot S_i^y(y). \tag{2.23}$$

The corresponding L^2-norm is then readily evaluated as follows

$$\|u^n\|_2 = \left(\int_{\Omega_x \times \Omega_y} \sum_{i=1}^{\frac{n \cdot (n+1)}{2}} S_i^x(x) \cdot S_i^y(y) \, dx \cdot dy \right)^{\frac{1}{2}} \tag{2.24}$$

$$= \left(\sum_{i=1}^{\frac{n \cdot (n+1)}{2}} \left(\int_{\Omega_x} S_i^x(x) \, dx \cdot \int_{\Omega_y} S_i^y(y) \, dy \right) \right)^{\frac{1}{2}}. \tag{2.25}$$

The above stopping criterion involves the evaluation of $2 + n \cdot (n + 1)$ one-dimensional integrals. A similar but less expensive criterion is based on the norm of mode n with respect to the norm of the first mode, i.e.

$$\mathcal{E}(n) = \frac{\|X_n(x) \cdot Y_n(y)\|}{\|X_1(x) \cdot Y_1(y)\|}. \tag{2.26}$$

This second criterion requires the computation of only two one-dimensional integrals.

More appropriate error estimators can be associated to the residual $R(n)$ obtained by inserting the PGD approximation $u^n(x, y)$ into the Poisson equation:

$$R(n) = \sum_{i=1}^{n} \left(\frac{\partial^2 X_i}{\partial x^2} \cdot Y_i(y) + X_i(x) \cdot \frac{\partial^2 Y_i}{\partial y^2} \right) - f. \tag{2.27}$$

Selecting $\mathcal{E}(n) = \|R(n)\|_2$ as error estimator again leads to the computation of one-dimensional integrals as with the previous criteria.

Other error estimators based on quantities of interest in the study of a particular problem are proposed in [1, 2].

2.1.3 Numerical Example

Let us consider the Poisson Eq. (2.1) on a two-dimensional rectangular domain $\Omega = \Omega_x \times \Omega_y = (0, 2) \times (0, 1)$ for which we seek a solution using the procedure described above. The source term f is set to $f = 1$, in which case we obtain analytically the following exact solution for $u(x, y)$:

$$u_{ex}(x, y) = \sum_{m,n \text{ odd}} \frac{64}{\pi^4 \left(4n^2 + m^2\right)} \cdot \sin\left(\frac{m\pi x}{2}\right) \cdot \sin\left(n\pi y\right). \qquad (2.28)$$

The unknown functions $X_i(x)$ and $Y_i(y)$ are sought on a uniform grid with M points. All one-dimensional differential problems arising in the solution procedure are solved using second-order finite differences, while the integrals are evaluated numerically using the trapezoidal rule. The chosen stopping criteria for the fixed point iterations and the enrichment process are those described by (2.8) and (2.26), respectively.

In Figs. 2.1 and 2.2 the normalized functions $X_i(x)$ and $Y_i(y)$ are illustrated for $i = 1, \ldots, 4$ and $M = 101$. One can observe that, as i increases, $X_i(x)$ and $Y_i(y)$ both account for a higher frequency content of the numerical solution.

In Fig. 2.3, we show the 2D reconstructed PGD solution together with the normalized functions $X_i(x)$ and $Y_i(y)$. For clarity reasons the 2D solution is shown on a coarser grid.

In the sequel we illustrate the convergence of the PGD solution towards the analytical solution as a function of both the discretization and the number of enrichment steps. For this purpose we define the following quadratic error between the analytical solution $u_{ex}(x, y)$ and a numerical PGD solution $u_M^N(x, y)$ with N enrichment steps and M discretization points for each coordinate:

Fig. 2.1 Normalized functions $X_i(x)$ for $i = 1, \ldots, 4$ produced by the PGD solution of (2.1)

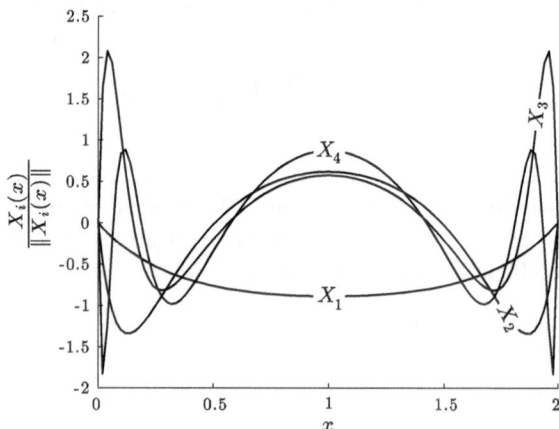

Fig. 2.2 Normalized functions $Y_i(y)$ for $i = 1, \ldots, 4$ produced by the PGD solution of (2.1)

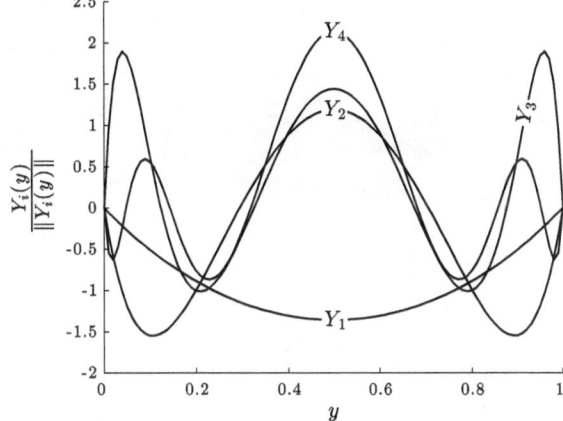

Fig. 2.3 Reconstructed 2D PGD solution of (2.1) and normalized functions $X_i(x)$ and $Y_i(y)$. The 2D solution is shown on a coarser grid

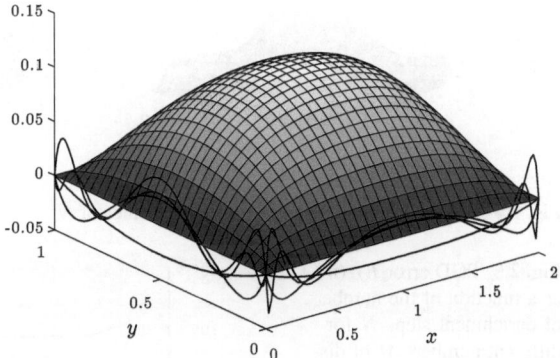

$$E_M(u_M^N) = \int_0^{\tilde{1}} \int_0^{\tilde{2}} \left(u_{\text{ex}}(x, y) - u_M^N(x, y) \right)^2 dx \cdot dy. \tag{2.29}$$

Here, the symbol $\tilde{\int}$ refers to a numerical integration carried out with the trapezoidal rule on the nodal values. We first illustrate the pointwise convergence of the PGD solution towards the exact solution in Fig. 2.4 where we plot $u_{\text{ex}} - u_M^N$ for $M = 41$ and different values of N. For clarity reasons the 2D error is shown on a coarser grid.

We further illustrate the convergence of the PGD solution by showing $E_M(u_M^N)$ as a function of N and as a function of M in Figs. 2.5 and 2.6 respectively. One can immediately notice that, for this problem, only a few enrichment steps are necessary for the PGD to converge to the analytical solution.

In the next example, we compute the first 10 enrichment steps on a coarse mesh ($M = 21$). We then interpolate the computed functions $X_i(x)$ and $Y_i(y)$ on a finer mesh ($M = 41$) using a natural spline interpolation, before computing another 10 enrichment steps on the fine mesh. The error $E_M(u_M^N)$ is shown in Fig. 2.7. This illustrates that after a few enrichment steps the error is mostly a discretization error.

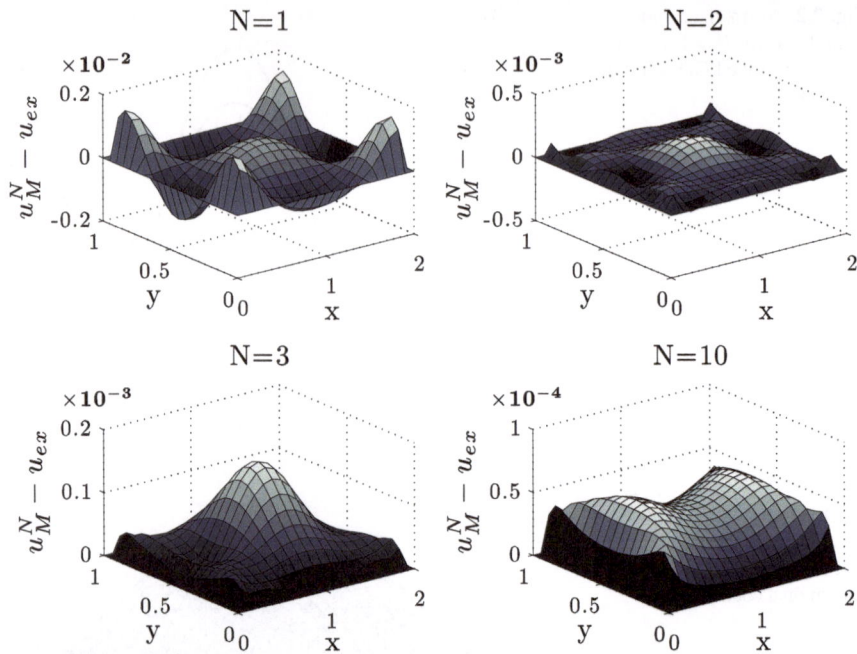

Fig. 2.4 $u_{\mathrm{ex}} - u_M^N$ for $M = 41$ and different values of N

Fig. 2.5 PGD error $E_M(u_M^N)$ as a function of the number of enrichment steps N for different numbers M of discretization points

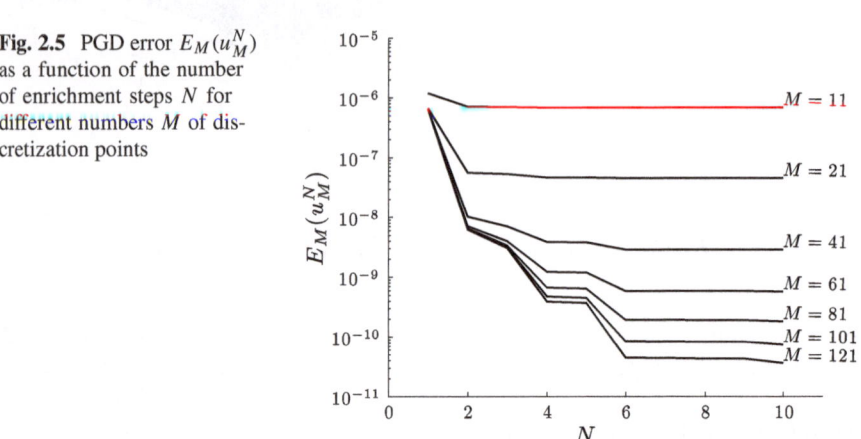

2.2 Taking into Account Neumann Boundary Conditions

In the sequel, we still consider the Poisson equation (2.1) with a constant source term defined in the domain $\Omega = \Omega_x \times \Omega_y$. The boundary conditions are somewhat different, however. We indeed specify a flux or Neumann condition along part of the domain boundary:

Fig. 2.6 PGD error $E_M(u_M^N)$ as a function of the number of grid points M for different numbers N of enrichment steps

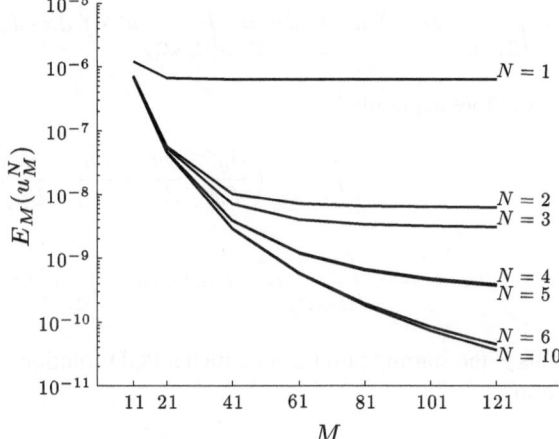

Fig. 2.7 PGD error $E_M(u_M^N)$ as a function of the number of enrichment steps. The first 10 enrichment steps have been computed on a coarse mesh with $M = 21$. The following enrichment steps have been computed on a finer mesh with $M = 41$

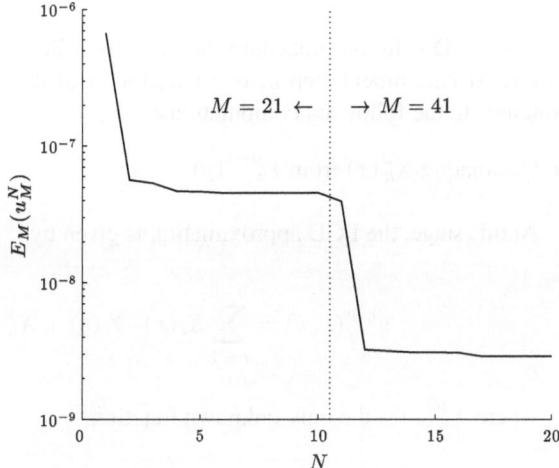

$$\begin{cases} u(x = 0, y) = 0 \\ u(x = L, y) = 0 \\ u(x, y = 0) = 0 \\ \dfrac{\partial u}{\partial y}\big|_{x, y=H} = q \end{cases} \qquad (2.30)$$

The classical way of accounting for Neumann conditions is to integrate by parts the weighted residual form (2.3) and implement the flux condition as a so-called natural boundary condition:

$$- \int_{\Omega_x \times \Omega_y} \nabla u^* \cdot \nabla u \, dx \cdot dy = \int_{\Omega_x \times \Omega_y} u^* \cdot f \, dx \cdot dy - \int_{\Omega_x} u^*(x, y = H) \cdot q \, dx, \tag{2.31}$$

or more explicitly

$$\int_{\Omega_x \times \Omega_y} \left(\frac{\partial u^*}{\partial x} \cdot \frac{\partial u}{\partial x} + \frac{\partial u^*}{\partial y} \cdot \frac{\partial u}{\partial y} \right) dx \cdot dy$$

$$= - \int_{\Omega_x \times \Omega_y} u^* \cdot f \, dx \cdot dy + \int_{\Omega_x} u^*(x, y = H) \cdot q \, dx. \tag{2.32}$$

This is the starting point from which a PGD solution can be sought in the separated form

$$u(x, y) = \sum_{i=1}^{N} X_i(x) \cdot Y_i(y). \tag{2.33}$$

The PGD solution procedure then readily follows as described in the first case study. At enrichment step n, one iteration p of the alternating direction strategy amounts to the following computations:

- Calculating $X_n^p(x)$ from $Y_n^{p-1}(y)$

At this stage, the PGD approximation is given by

$$u^{n,p}(x, y) = \sum_{i=1}^{n-1} X_i(x) \cdot Y_i(y) + X_n^p(x) \cdot Y_n^{p-1}(y), \tag{2.34}$$

where $X_n^p(x)$ is the only unknown function.

Using Galerkin's method, we select the following weight function

$$u^*(x, y) = X_n^*(x) \cdot Y_n^{p-1}(y). \tag{2.35}$$

Inserting (2.34) and (2.35) into (2.32), we obtain

$$\int_{\Omega_x \times \Omega_y} \left(\frac{dX_n^*}{dx} \cdot \frac{dX_n^p}{dx} \cdot \left(Y_n^{p-1} \right)^2 + X_n^* \cdot X_n^p \cdot \left(\frac{dY_n^{p-1}}{dy} \right)^2 \right) dx \cdot dy$$

$$= - \int_{\Omega_x \times \Omega_y} \sum_{i=1}^{n-1} \left(\frac{dX_n^*}{dx} \cdot \frac{dX_i}{dx} \cdot Y_n^{p-1} \cdot Y_i + X_n^* \cdot X_i \cdot \frac{dY_n^{p-1}}{dy} \cdot \frac{dY_i}{dy} \right) dx \cdot dy$$

$$-\int_{\Omega_x \times \Omega_y} X_n^* \cdot Y_n^{p-1} \cdot f \, dx \cdot dy + \int_{\Omega_x} X_n^* \cdot Y_n^{p-1}(y = H) \cdot q \, dx. \quad (2.36)$$

In the above expression, all functions of the coordinate y are known, and we can evaluate the corresponding one-dimensional integrals:

$$\begin{cases} \alpha^x = \int_{\Omega_y} \left(Y_n^{p-1}(y) \right)^2 dy \\ \beta^x = \int_{\Omega_y} \left(\dfrac{dY_n^{p-1}(y)}{dy} \right)^2 dy \\ \gamma_i^x = \int_{\Omega_y} Y_n^{p-1}(y) \cdot Y_i(y) \, dy \\ \delta_i^x = \int_{\Omega_y} \dfrac{dY_n^{p-1}(y)}{dy} \cdot \dfrac{dY_i(y)}{dy} \, dy \\ \xi^x = \int_{\Omega_y} Y_n^{p-1}(y) \cdot f \, dy \\ \mu^x = Y_n^{p-1}(y = H) \cdot q \end{cases} \quad (2.37)$$

We thus obtain the weighted residual form of an elliptic problem for $X_n^p(x)$ defined over Ω_x:

$$\int_{\Omega_x} \left(\frac{dX_n^*}{dx} \cdot \frac{dX_n^p}{dx} \cdot \alpha^x + X_n^* \cdot X_n^p \cdot \beta^x \right) dx$$

$$= -\int_{\Omega_x} \sum_{i=1}^{n-1} \left(\frac{dX_n^*}{dx} \cdot \frac{dX_i}{dx} \cdot \gamma_i^x + X_n^* \cdot X_i \cdot \delta_i^x \right) dx$$

$$-\int_{\Omega_x} X_n^* \cdot \xi^x \, dx + \int_{\Omega_x} X_n^* \cdot \mu^x \, dx. \quad (2.38)$$

The finite element method, for example, can then be used to discretize this one-dimensional problem, with the remaining Dirichlet condition $X_n^p(x = 0) = X_n^p(x = L) = 0$.

- Calculating $Y_n^p(y)$ from the just computed $X_n^p(x)$

Here again, the second step of iteration p simply mirrors the first one with an exchange of role between x and y coordinates.

The current PGD approximation reads

$$u^{n,p}(x, y) = \sum_{i=1}^{n-1} X_i(x) \cdot Y_i(y) + X_n^p(x) \cdot Y_n^p(y), \quad (2.39)$$

where $Y_n^p(y)$ is the only unknown function.

Selecting the Galerkin method, i.e.

$$u^*(x, y) = X_n^P(x) \cdot Y_n^*(y),$$ (2.40)

we introduce (2.39) and (2.40) into (2.32) to obtain

$$\int_{\Omega_x \times \Omega_y} \left(\left(\frac{dX_n^P}{dx} \right)^2 \cdot Y_n^* \cdot Y_n^P + (X_n^P)^2 \cdot \frac{dY_n^*}{dy} \cdot \frac{dY_n^P}{dy} \right) dx \cdot dy$$

$$= -\int_{\Omega_x \times \Omega_y} \sum_{i=1}^{n-1} \left(\frac{dX_n^P}{dx} \cdot \frac{dX_i}{dx} \cdot Y_n^* \cdot Y_i + X_n^P \cdot X_i \cdot \frac{dY_n^*}{dy} \cdot \frac{dY_i}{dy} \right) dx \cdot dy$$

$$- \int_{\Omega_x \times \Omega_y} X_n^P \cdot Y_n^* \cdot f \, dx \cdot dy + \int_{\Omega_x} X_n^P \cdot Y_n^*(y = H) \cdot q \, dx.$$ (2.41)

Now, all functions of x are known, and we can compute the integrals

$$\begin{cases} \alpha^y = \int_{\Omega_x} \left(X_n^P(x) \right)^2 dx \\ \beta^y = \int_{\Omega_x} \left(\frac{dX_n^P(x)}{dx} \right)^2 dx \\ \gamma_i^y = \int_{\Omega_x} X_n^P(x) \cdot X_i(x) \, dx \\ \delta_i^y = \int_{\Omega_x} \frac{dX_n^P(x)}{dx} \cdot \frac{dX_i(x)}{dx} \, dx \\ \xi^y = \int_{\Omega_x} X_n^P(x) \cdot f \, dx \\ \mu^y = \int_{\Omega_x} X_n^P(x) \cdot q \, dx \end{cases}$$ (2.42)

We thus obtain the weighted residual form of an elliptic problem for $Y_n^P(y)$ defined over Ω_y:

$$\int_{\Omega_y} \left(\beta^y \cdot Y_n^* \cdot Y_n^P + \alpha^y \cdot \frac{dY_n^*}{dy} \cdot \frac{dY_n^P}{dy} \right) dy$$

$$= -\int_{\Omega_y} \sum_{i=1}^{n-1} \left(\delta_i^y \cdot Y_n^* \cdot Y_i + \gamma_i^y \cdot \frac{dY_n^*}{dy} \cdot \frac{dY_i}{dy} \right) dy$$

$$- \int_{\Omega_y} \xi^y \cdot Y_n^* \, dy + Y_n^*(y = H) \cdot \mu^y.$$ (2.43)

Here again, we can use the finite element method to discretize this one-dimensional problem, with the remaining Dirichlet conditions $Y_n^P(y = 0) = 0$.

2.3 Increasing the Complexity of the Case Study

2.3.1 Non-Constant Source Term

We have assumed so far a constant source term f. We now extend the PGD strategy to the case of a non-uniform source $f(x, y)$. We shall see in Sect. 3.2 how to obtain a separated representation of f in the form

$$f(x, y) = \sum_{j=1}^{\mathcal{F}} F_j^x(x) \cdot F_j^y(y). \tag{2.44}$$

With the following notation,

$$\xi_j^x = \int_{\Omega_y} Y_n^{p-1}(y) \cdot F_j^y(y) \, dy, \tag{2.45}$$

it is then easy to verify that (2.13) and (2.38) become respectively

$$\int_{\Omega_x} X_n^* \cdot \left(\alpha^x \cdot \frac{d^2 X_n^p}{dx^2} + \beta^x \cdot X_n^p \right) dx$$

$$= - \int_{\Omega_x} X_n^* \cdot \sum_{i=1}^{n-1} \left(\gamma_i^x \cdot \frac{d^2 X_i}{dx^2} + \delta_i^x \cdot X_i \right) dx + \int_{\Omega_x} X_n^* \cdot \left(\sum_{j=1}^{\mathcal{F}} \xi_j^x \cdot F_j^x(x) \right) dx, \tag{2.46}$$

and

$$\int_{\Omega_x} \left(\frac{dX_n^*}{dx} \cdot \frac{dX_n^p}{dx} \cdot \alpha^x + X_n^* \cdot X_n^p \cdot \beta^x \right) dx$$

$$= - \int_{\Omega_x} \sum_{i=1}^{n-1} \left(\frac{dX_n^*}{dx} \cdot \frac{dX_i}{dx} \cdot \gamma_i^x + X_n^* \cdot X_i \cdot \delta_i^x \right) dx$$

$$- \int_{\Omega_x} X_n^* \cdot \left(\sum_{j=1}^{\mathcal{F}} \xi_j^x \cdot F_j^x(x) \right) dx + \int_{\Omega_x} X_n^* \cdot \mu^x \, dx. \tag{2.47}$$

Similarly, with the definition

$$\xi_j^y = \int_{\Omega_x} X_n^p(x) \cdot F_j^x(x) \, dx, \tag{2.48}$$

Equations (2.19) and (2.43) become respectively

$$\int_{\Omega_y} Y_n^* \cdot \left(\alpha^y \cdot \frac{d^2 Y_n^p}{dy^2} + \beta^y \cdot Y_n^p \right) dy$$

$$= -\int_{\Omega_y} Y_n^* \cdot \sum_{i=1}^{n-1} \left(\gamma_i^y \cdot \frac{d^2 Y_i}{dy^2} + \delta_i^y \cdot Y_i \right) dy + \int_{\Omega_y} Y_n^* \cdot \left(\sum_{j=1}^{j=\mathcal{F}} \xi_j^y \cdot F_j^y(y) \right) dy$$

$$(2.49)$$

and

$$\int_{\Omega_y} \left(\beta^y \cdot Y_n^* \cdot Y_n^p + \alpha^y \cdot \frac{dY_n^*}{dy} \cdot \frac{dY_n^p}{dy} \right) dy$$

$$= -\int_{\Omega_y} \sum_{i=1}^{n-1} \left(\delta_i^y \cdot Y_n^* \cdot Y_i + \gamma_i^y \cdot \frac{dY_n^*}{dy} \cdot \frac{dY_i}{dy} \right) dy$$

$$- \int_{\Omega_y} Y_n^* \cdot \left(\sum_{j=1}^{j=\mathcal{F}} \xi_j^y \cdot F_j^y(y) \right) dy + Y_n^*(y = H) \cdot \mu^y. \qquad (2.50)$$

The same procedure is used when the problem to be solved has non-constant coefficients.

2.3.2 Non-Homogeneous Dirichlet Boundary Conditions

Let us now specify non-homogeneous Dirichlet conditions along a part Γ_D of the domain boundary Γ: $u(x, y) = \bar{u}(x, y) \neq 0$ for $(x, y) \in \Gamma_D$.

In order to apply the PGD strategy, we simply consider a function $g(x, y)$ regular enough that satisfies the same Dirichlet conditions, i.e. $g(x, y) = \bar{u}(x, y)$ for $(x, y) \in \Gamma_D$, but is otherwise arbitrary [3].

Then again, as explained in Sect. 3.2, we compute *a priori* the separated representation of the function $g(x, y)$

$$g(x, y) = \sum_{j=1}^{\mathcal{G}} G_j^x(x) \cdot G_j^y(y). \qquad (2.51)$$

This expansion can be seen as a (very) approximate solution that satisfies the Dirichlet boundary conditions but does not verify neither the partial differential equation nor the natural boundary conditions. In order to enforce both, it suffices to enrich this approximation to obtain

$$u(x, y) = \sum_{j=1}^{N} X_i(x) \cdot Y_i(y), \tag{2.52}$$

where $X_i(x) = G_i^x(x)$ and $Y_i(y) = G_i^y(y)$, for $i = 1, \ldots, \mathcal{G}$. The remaining functions $X_i(x)$ and $Y_i(y)$, for $i > \mathcal{G}$, are calculated by using the PGD procedure described previously for homogeneous Dirichlet conditions.

2.3.3 Higher Dimensions and Separability of the Computational Domain

The PGD procedure described in this chapter can easily be generalized to models defined in D dimensions as long as the computational domain is *separable*. By this we mean that the domain is the Cartesian product of one-dimensional intervals:

$$\Omega = \Omega_1 \times \Omega_2 \times \cdots \times \Omega_D. \tag{2.53}$$

Thus, the unknown field $u(x_1, \ldots, x_D)$ is sought in the separated form

$$u(x_1, \ldots, x_D) = \sum_{i=1}^{N} X_i^1(x_1) \times \cdots \times X_i^D(x_D). \tag{2.54}$$

For example, should we consider the Poisson equation, then its PGD solution in Ω would involve a series of decoupled one-dimensional problems (second-order ordinary differential equations) formulated in each Ω_i.

For a non-separable domain, one possible approach consists in embedding Ω into a separable domain $\overline{\Omega} = \overline{\Omega}_1 \times \overline{\Omega}_2 \times \cdots \times \overline{\Omega}_D$ such that $\Omega \subset \overline{\Omega}$. One then applies the PGD strategy in $\overline{\Omega}$ together with an appropriate penalty formulation in $\overline{\Omega} \setminus \Omega$ [3].

In many applications, the computational domain is not strictly separable according to our definition, but it is the Cartesian product of multi-dimensional sets. One can then easily apply a PGD strategy by separating the coordinates into the several groups that correspond to these sets. For example, consider a three-dimensional extruded domain $\Omega = \Omega_{(x,y)} \times \Omega_z$. Here, the domain Ω is the extrusion of the non-separable two-dimensional domain $\Omega_{(x,y)}$ along the z axis. In this case, the PGD decomposition reads most naturally

$$u(x, y, z) = \sum_{i=1}^{N} X_i(x, y) \cdot Z_i(z). \tag{2.55}$$

This approach is particularly well suited to problems involving plates, shells or profiled geometries. The PGD calculations thus involve a series of decoupled two-dimensional problems in $\Omega_{(x,y)}$ to compute the functions $X_i(x, y)$, and one-dimensional problems in Ω_z to compute the functions $Z_i(z)$. As a result, the fully-three dimensional PGD simulation has a numerical complexity typical of two-dimensional analyses.

2.4 Numerical Examples

2.4.1 2D Heat Transfer Problem

In this section, we illustrate the developments of the previous sections by comparing the PGD and the finite element solutions of the following problem:

$$\Delta u(x, y) = f(x, y), \tag{2.56}$$

defined in a two-dimensional rectangular domain $\Omega = \Omega_x \times \Omega_y = (0, 2) \times (0, 1)$. The boundary conditions are specified as follows:

$$\begin{cases} u(x = 0, y) = y \cdot (1 - y) \\ u(x = 2, y) = 0 \\ u(x, y = 0) = 0 \\ \dfrac{\partial u}{\partial y}\big|_{x, y = 1} = -1 \end{cases} \tag{2.57}$$

Thus, we can write (2.51) as:

$$g(x, y) = G_1^x(x) \cdot G_1^y(y) = \frac{2}{2} \frac{x}{} \cdot y \cdot (1 - y). \tag{2.58}$$

Finally, we consider the following separated representation for the source term $f(x, y)$:

$$f(x, y) = F_1^x(x) \cdot F_1^y(y) = -5e^{-10 \cdot (x-1)^2} \cdot e^{-10 \cdot (y-0.5)^2}. \tag{2.59}$$

The unknown functions $X_i(x)$ and $Y_i(y)$ are sought on a uniform grid with $M = 41$ points. All one-dimensional differential problems and integrals arising in the solution procedure are solved or computed using linear one-dimensional finite elements. The stopping criteria for the fixed point iterations and the enrichment process are those described by (2.8) and (2.26), respectively. Error levels are computed using the following expression:

$$E_M(u_M^N) = \int_0^1 \int_0^2 \left(u_{FE, M}(x, y) - u_M^N(x, y) \right)^2 dx \cdot dy, \tag{2.60}$$

where $u_{FE, M}(x, y)$ is the corresponding 2D finite element solution on an equivalent mesh.

Fig. 2.8 Normalized functions $G_1^x(x)$ and $X_i(x)$ for $i = 1, \ldots, 4$

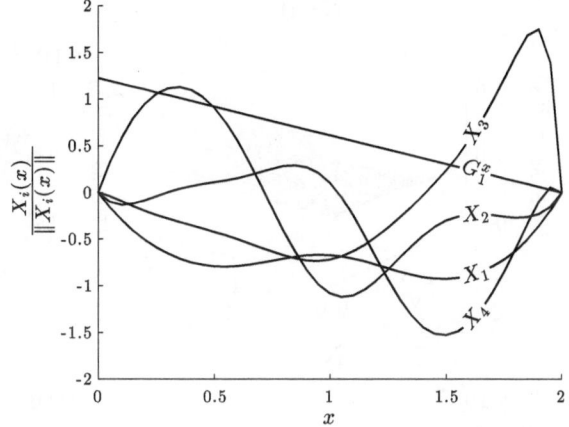

Fig. 2.9 Normalized functions $G_1^y(y)$ and $Y_i(y)$ for $i = 1, \ldots, 4$

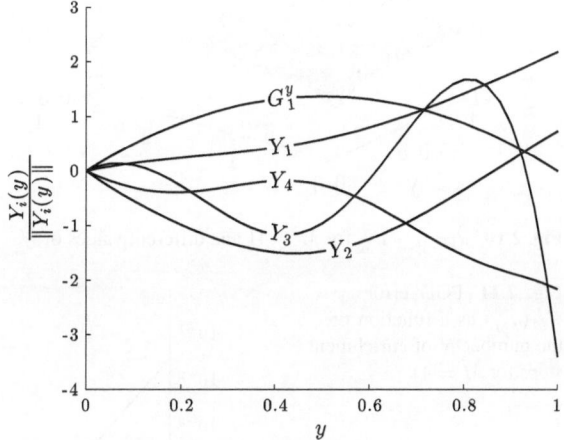

In Figs. 2.8 and 2.9, we show the normalized functions $G_1^x(x)$, $X_i(x)$, $G_1^y(y)$ and $Y_i(y)$ for $i = 1, \ldots, 4$. In Fig. 2.8 one can observe that only $G_1^x(x)$ is non-zero for $x = 0$ since all the $X_i(x)$ have to preserve the non-zero Dirichlet condition imposed through $G_1^x(x)$ and $G_1^y(y)$.

We first illustrate the pointwise convergence of the PGD solution towards the reference solution in Fig. 2.10 where we plot $u_{\text{FE},M} - u_M^N$ for $M = 41$ and different values of N. For $N = 0$, the solution reduces to $u(x, y) = g(x, y)$. We further illustrate the convergence of the PGD solution in Fig. 2.11 by showing $E_M(u_M^N)$ as a function of N.

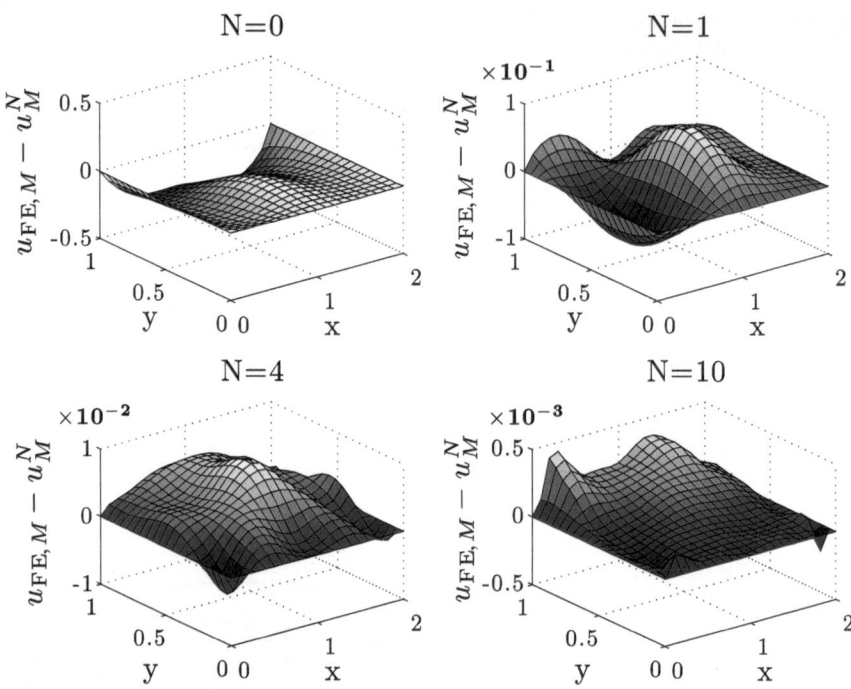

Fig. 2.10 $u_{\text{FE},M} - u_M^N$ for $M = 41$ and different values of N

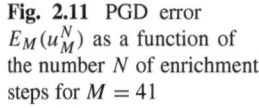

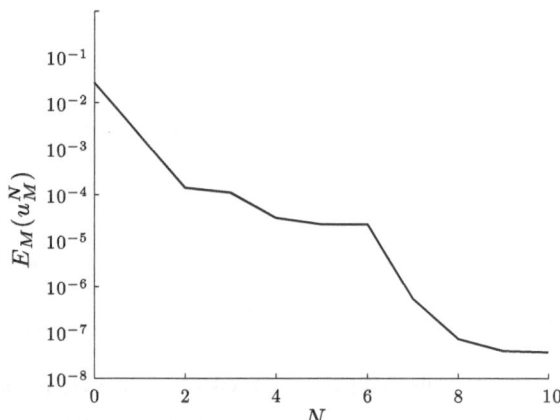

Fig. 2.11 PGD error $E_M(u_M^N)$ as a function of the number N of enrichment steps for $M = 41$

2.4.2 High-Dimensional Problem

We now focus on the solution of the following high-dimensional problem:

$$\Delta u(x_1, \ldots, x_D) = f(x_1, \ldots, x_D), \tag{2.61}$$

in a D-dimensional hypercube $\Omega = \Omega_1 \times \cdots \times \Omega_D = (-1, 1) \times \cdots \times (-1, 1)$. The boundary conditions are homogeneous on the whole boundary of Ω. The source term $f(x_1, \ldots, x_D)$ is taken such that we have the following 2-term separated solution:

$$u_{\text{ex}}^D = \prod_{d=1}^{D} x_d \cdot \sin(d \cdot \pi \cdot x_d) + \prod_{d=1}^{D} x_d^2 \cdot \sin((D + 1 - d) \cdot \pi \cdot x_d). \qquad (2.62)$$

Similarly to the previous examples, we seek a N-term PGD solution of the form:

$$u_{M,N}^D = \sum_{i=1}^{N} \prod_{d=1}^{D} X_i^d(x_d), \qquad (2.63)$$

where M is the number of finite element nodal values along each dimension. Remember that, using standard grid-based methods, the solution would involve M^D degrees of freedom. Again, the one-dimensional integrals and BVP arising in the PGD solution procedure are computed using linear finite elements. Error levels are computed using the following expression:

$$E_M^D(u_{M,N}^D) = \int_{\Omega_1}^{\tilde{}} \cdots \int_{\Omega_D}^{\tilde{}} \left(u_{\text{ex}}^D - u_{M,N}^D \right)^2 dx_1 \cdots dx_D, \qquad (2.64)$$

where $\tilde{\int}$ refers to the numerical integration on the finite element mesh. We show in Fig. 2.12, for $M = 101$, the decrease of the normalized error levels for different values of D as we increase the number of terms in the PGD solution. In this particular case, the PGD solution is optimal in the sense that after only two enrichment steps, the relative error is about $\cdot 10^{-15}$ and does not decrease upon further enrichment.

Fig. 2.12 PGD error $E_M^D(u_{M,N}^D)$ as a function of the number N of enrichment steps for different values of D and $M = 101$

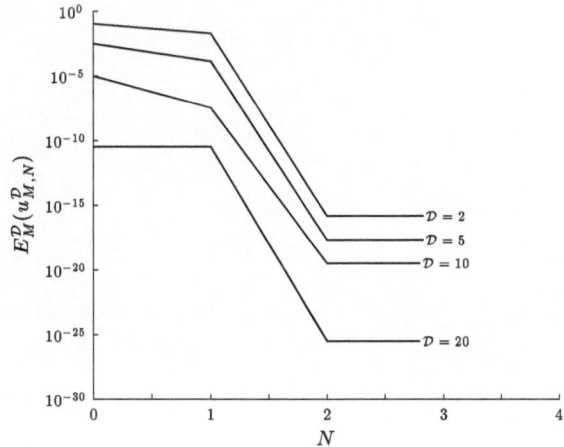

For $D = 10$, the PGD solution is computed in a few seconds on a laptop computer. Traditional grid-based methods with the same level of discretization would require more than 10^{20} degrees of freedom, which is far beyond the capabilities of today's computers.

References

1. A. Ammar, F. Chinesta, P. Diez, A. Huerta, An error estimator for separated representations of highly multidimensional models. Comput. Methods Appl. Mech. Eng. **199**, 1872–1880 (2010)
2. P. Ladevèze, L. Chamoin, On the verification of model reduction methods based on the proper generalized decomposition. Comput. Methods Appl. Mech. Eng. **200**, 2032–2047 (2011)
3. D. Gonzalez, A. Ammar, F. Chinesta, E. Cueto, Recent advances in the use of separated representations. Int. J. Numer. Meth. Eng. **81/5**, 637–659 (2010)

Chapter 3
PGD Versus SVD

Abstract The issue of separability is of major importance when using the Proper Generalized Decomposition. Efficient computer implementations require the separated representation of model parameters, boundary conditions and/or source terms. This can be performed by applying the Singular Value Decomposition (SVD) or its multi-dimensional counterpart, the so-called High Order Singular Value Decomposition. This chapter revisits these concepts. We then point out and discuss the subtle connections between SVD and PGD. This allows us to illustrate how the PGD solver can compress separated representations, and also to justify the fact that in certain circumstances the PGD solution procedure can be viewed as the calculation of an on-the-fly compressed representation.

Keywords Data Compression · Proper Generalized Decomposition · Separated representation · Singular Value Decomposition

3.1 Singular Value Decomposition and On-the-Fly Data Compression Using the PGD

Before discussing more complex applications, it is useful to pause for a while and reflect on the efficiency of the PGD in the context of the classical data compression problem.

Consider the numerical approximation of a two-dimensional field $f(x, y)$ computed by means of a standard mesh-based technique. The discrete solution is given by the set of nodal values $f(x_j, y_i)$ with $i = 1, \ldots, \mathcal{I}$ and $j = 1, \ldots, \mathcal{J}$. The representation of the solution requires $\mathcal{I} \times \mathcal{J}$ data. This can be a large amount indeed, and even more so for higher-dimensional problems. The question then is how to efficiently compress these data.

A well-known approach to data compression relies on the Singular Value Decomposition (SVD). Let $\mathbf{M}$ be a $\mathcal{I} \times \mathcal{J}$ matrix of rank r and entries $M_{ij} = f(x_j, y_i)$. As

F. Chinesta et al., *The Proper Generalized Decomposition for Advanced Numerical Simulations*, SpringerBriefs in Applied Sciences and Technology, DOI: 10.1007/978-3-319-02865-1_3, © The Author(s) 2014

for any real matrix, one can apply the SVD to factorize $\mathbf{M}$ in the form

$$\mathbf{M} = \mathbf{U} \cdot \Sigma \cdot \mathbf{V}^T. \tag{3.1}$$

Here, $\mathbf{U}$ is a $\mathcal{I} \times \mathcal{I}$ real unitary matrix, i.e. $\mathbf{U} \cdot \mathbf{U}^T$ is the identity matrix. The columns of $\mathbf{U}$ are the eigenvectors of $\mathbf{M} \cdot \mathbf{M}^T$, also known as the left-singular eigenvectors of $\mathbf{M}$. Similarly, $\mathbf{V}$ is a $\mathcal{J} \times \mathcal{J}$ real unitary matrix, and its columns are the eigenvectors of $\mathbf{M}^T \cdot \mathbf{M}$, or the right-singular eigenvectors of $\mathbf{M}$. The matrix Σ is an $\mathcal{I} \times \mathcal{J}$ rectangular diagonal matrix, with r positive real entries σ_i on the diagonal and zeros elsewhere. The σ_i's are the singular values of $\mathbf{M}$, ordered such that σ_1 is the largest one.

Now let $\mathbf{U}_i$ and $\mathbf{V}_i$ denote the columns of the corresponding SVD matrices $\mathbf{U}$ and $\mathbf{V}$. We define the separable matrix $\mathbf{M}_i$ as the scaled dot product $\mathbf{M}_i = \sigma_i \, \mathbf{U}_i \cdot \mathbf{V}_i^T$, or $(\mathbf{M}_i)_{kl} = \sigma_i U_{ki} V_{li}$. The SVD (3.1) can then be seen as the decomposition of a matrix into a weighted, ordered sum of separable matrices:

$$\mathbf{M} = \sum_{i=1}^{r} \mathbf{M}_i = \sum_{i=1}^{r} \sigma_i \cdot \mathbf{U}_i \cdot \mathbf{V}_i^T. \tag{3.2}$$

This separated representation usually allows for significant data compression. Indeed, if one only keeps the N first terms in (3.2) corresponding to the N largest singular values ($N \ll r$), the solution is suitably represented by N vectors of size $\mathcal{I}$ and N vectors of size $\mathcal{J}$. The total amount of data is thus $N \cdot (\mathcal{I} + \mathcal{J})$, which is typically much less than $\mathcal{I} \times \mathcal{J}$.

In view of the above discussion, it becomes clear that the PGD can be viewed as a solver that computes *a priori* or *on the fly* a compressed representation of the solution. For symmetric and positive-definite differential operators, we noticed [1] that the Nth-order PGD approximation (see also [2, 3])

$$f^N(x, y) = \sum_{i=1}^{N} F_i^x(x) \cdot F_i^y(y), \tag{3.3}$$

is very close to the separated representation computed by applying the SVD on the model solution and selecting the N first terms related to the highest singular values:

$$\left\| \mathbf{M} - \sum_{i=1}^{N} \sigma_i \cdot \mathbf{U}_i \cdot \mathbf{V}_i^T \right\| \approx \left\| \mathbf{M} - \sum_{i=1}^{N} \mathbf{F}_i^y \cdot \mathbf{F}_i^{xT} \right\|. \tag{3.4}$$

In the above expression, the vectors $\mathbf{F}_i^x$ and $\mathbf{F}_i^y$ are formed by evaluating the functions $F_i^x(x)$ and $F_i^y(y)$ at discrete points x_k and y_l, respectively. The matrix

$$\mathbf{F}^N = \sum_{i=1}^{N} \mathbf{F}_i^y \cdot \mathbf{F}_i^{xT}, \tag{3.5}$$

is thus the discrete reconstructed form of the PGD solution.

For more general differential operators, the separated representation resulting from the application of the PGD is not necessarily optimal. As we shall detail later, a data post-compression step can then be applied to the PGD solution if deemed necessary.

The PGD solver computes on the fly a compressed solution even for models defined in high-dimensional spaces. It is important to stress that the multidimensional extension of the SVD, known as the high-order singular value decomposition (HOSVD), is only optimal for two-dimensional matrices. Moreover, even if HOSVD strategies exist, their main limitation concerns the storage and manipulation of multidimensional arrays [4].

3.2 The PGD for Constructing Separated Approximations

We now analyze the connexion between the PGD approximation constructor and the SVD. In view of the discussion in Sect. 2.3, we use the PGD for calculating the N-term separated approximation $u^N(x, y)$ of a given function $f(x, y)$. Thus, we wish to compute an approximation of $u(x, y)$ governed by the following algebraic problem

$$u(x, y) = f(x, y), \quad (x, y) \in \Omega = \Omega_x \times \Omega_y. \tag{3.6}$$

The corresponding weighted residual form reads

$$\int_{\Omega_x \times \Omega_y} u^* \cdot (u(x, y) - f(x, y)) \, dx \cdot dy = 0, \quad \forall u^* \tag{3.7}$$

In order to emphasize the connexion between PGD and SVD, we begin our analysis with the one-term PGD approximation

$$u^1(x, y) = X(x) \cdot Y(y). \tag{3.8}$$

As is now customary, the resulting non-linear problem is solved iteratively by means of an alternating direction scheme. At each iteration, one first compute X from Y by solving

$$\int_{\Omega_x \times \Omega_y} X^* \cdot Y \cdot (X \cdot Y - f(x, y)) \, dx \cdot dy = 0, \tag{3.9}$$

and then compute Y from X,

$$\int_{\Omega_x \times \Omega_y} X \cdot Y^* \cdot (X \cdot Y - f(x, y)) \, dx \cdot dy = 0. \tag{3.10}$$

The strong forms of (3.9) and (3.10) thus yield

$$X = \frac{\int_{\Omega_y} Y \cdot f \, dy}{\int_{\Omega_y} Y^2 dy}, \tag{3.11}$$

and

$$Y = \frac{\int_{\Omega_x} X \cdot f \, dx}{\int_{\Omega_x} X^2 dx}. \tag{3.12}$$

Consider now the discrete analogs of (3.11) and (3.12) that involve the vector form of functions X and Y given by the values of these functions at points x_k and y_l, respectively. For the sake of simplicity, these points are assumed uniformly distributed in their respective intervals. Via numerical integration, we have

$$\mathbf{X} = \frac{\mathbf{M}^T \cdot \mathbf{Y}}{\mathbf{Y}^T \cdot \mathbf{Y}}, \tag{3.13}$$

and

$$\mathbf{Y} = \frac{\mathbf{M} \cdot \mathbf{X}}{\mathbf{X}^T \cdot \mathbf{X}}, \tag{3.14}$$

respectively, where $\mathbf{M}$ is the matrix form of $f(x, y)$ and $\mathbf{Y}^T \cdot \mathbf{Y}$ and $\mathbf{X}^T \cdot \mathbf{X}$ are scalars.

It then immediately follows from (3.13) and (3.14) that $\mathbf{X}$ and $\mathbf{Y}$ are eigenvectors of $\mathbf{M}^T \cdot \mathbf{M}$ and $\mathbf{M} \cdot \mathbf{M}^T$, respectively. These coincide with the left- and right-singular vectors making up the matrices $\mathbf{U}$ and $\mathbf{V}$ in the SVD decomposition (3.1) of $\mathbf{M}$.

Thus, in the two-dimensional case, SVD and PGD both yield an optimal separated representation of a given function. For dimensions strictly greater than two, HOSVD as well as PGD produce compact separated representations of a given function, but their optimality is not guaranteed.

3.3 The PGD Approximation Constructor in Action

We now go back to the algebraic approximation problem (3.6) and proceed to the computation of its N-term PGD solution

$$u^N(x, y) = \sum_{i=1}^{N} X_i(x) \cdot Y_i(y). \tag{3.15}$$

The procedure follows as explained in the previous chapters. At enrichment step n, we wish to compute

$$u^n(x, y) = u^{n-1}(x, y) + X_n(x) \cdot Y_n(y) = \sum_{i=1}^{n-1} X_i(x) \cdot Y_i(y) + X_n(x) \cdot Y_n(y), \quad (3.16)$$

wherein the functions $X_n(x)$ and $Y_n(y)$ are the only unknowns. An alternating direction scheme is used to solve this non-linear problem. At iteration p, this two-step scheme thus proceeds as follows:

- Calculating $X_n^p(x)$ from $Y_n^{p-1}(y)$

  The current approximation reads

$$u^n(x, y) = \sum_{i=1}^{n-1} X_i(x) \cdot Y_i(y) + X_n^p(x) \cdot Y_n^{p-1}(y), \quad (3.17)$$

where all functions except $X_n^p(x)$ are known.

The simplest choice for the weight function $u^*(x, y)$ in the weighted residual form (3.7) is

$$u^*(x, y) = X_n^*(x) \cdot Y_n^{p-1}(y). \quad (3.18)$$

Injecting (3.17) and (3.18) into (3.7), we obtain

$$\int_{\Omega_x \times \Omega_y} X_n^* \cdot Y_n^{p-1} \cdot \left(X_n^p \cdot Y_n^{p-1} \right) dx \cdot dy =$$

$$\int_{\Omega_x \times \Omega_y} X_n^* \cdot Y_n^{p-1} \cdot \left(f - \sum_{i=1}^{n-1} X_i \cdot Y_i \right) dx \cdot dy. \quad (3.19)$$

The y-integrals are then performed over Ω_y to yield

$$\int_{\Omega_x} \alpha^x \cdot X_n^* \cdot X_n^p \, dx = \int_{\Omega_x} X_n^* \cdot \left(\beta^x - \sum_{i=1}^{n-1} \gamma_i^x \cdot X_i \right) dx, \quad (3.20)$$

where

$$\begin{cases} \alpha^x = \int_{\Omega_y} \left(Y_n^{p-1}(y) \right)^2 dy \\ \beta^x(x) = \int_{\Omega_y} Y_n^{p-1}(y) \cdot f(x, y) \, dy \\ \gamma_i^x = \int_{\Omega_y} Y_n^{p-1}(y) \cdot Y_i(y) \, dy \end{cases} \quad (3.21)$$

The corresponding strong form reads

$$\alpha^x \cdot X_n^P(x) = \beta^x(x) - \sum_{i=1}^{n-1} \gamma_i^x \cdot X_i(x). \tag{3.22}$$

This defines an algebraic problem for the unknown function $X_n^P(x)$. Obviously, since the original approximation problem (3.6) does not involve derivatives of the unknown field $u(x, y)$, the equation defining $X_n^P(x)$ is also purely algebraic.

- Calculating $Y_n^P(y)$ from the just-computed $X_n^P(x)$

The current approximation now reads

$$u^n(x, y) = \sum_{i=1}^{n-1} X_i(x) \cdot Y_i(y) + X_n^P(x) \cdot Y_n^P(y), \tag{3.23}$$

where $Y_n^P(y)$ is the single unknown.

In the weighted residual form (3.7), we select the following weight function

$$u^*(x, y) = X_n^P(x) \cdot Y_n^*(y), \tag{3.24}$$

to obtain

$$\int_{\Omega_x \times \Omega_y} X_n^P \cdot Y_n^* \cdot \left(X_n^P \cdot Y_n^P \right) dx \cdot dy =$$

$$\int_{\Omega_x \times \Omega_y} X_n^P \cdot Y_n^* \cdot \left(f - \sum_{i=1}^{n-1} X_i \cdot Y_i \right) dx \cdot dy. \tag{3.25}$$

By integrating over Ω_x, we have

$$\int_{\Omega_y} \alpha^y \cdot Y_n^* \cdot Y_n^P \, dy = \int_{\Omega_y} Y_n^* \cdot \left(\beta^y - \sum_{i=1}^{n-1} \gamma_i^y \cdot Y_i \right) dy, \tag{3.26}$$

where

$$\begin{cases} \alpha^y = \int_{\Omega_x} \left(X_n^P(x) \right)^2 dx \\ \beta^y(y) = \int_{\Omega_x} X_n^P(x) \cdot f(x, y) \, dx \\ \gamma_i^y = \int_{\Omega_x} X_n^P(x) \cdot X_i(x) \, dx \end{cases} . \tag{3.27}$$

Finally, the corresponding strong form reads

$$\alpha^y \cdot Y_n^P(y) = \beta^y(y) - \sum_{i=1}^{n-1} \gamma_i^y \cdot Y_i(y). \tag{3.28}$$

This again is an algebraic problem for the unknown function $Y_n^P(y)$.

There are two issues in the application of this procedure, which become important when the dimensionality of the problem increases. One lies in the calculation of $\beta^x(x)$ and $\beta^y(y)$. With the functions X_i known at discrete positions x_k, this implies the evaluation of a large number of integrals

$$\beta^x(x_k) = \int_{\Omega_y} Y_n^{p-1}(y) \cdot f(x_k, y)\, dy, \quad k = 1, \ldots, \mathcal{J}, \tag{3.29}$$

and similarly for $\beta^y(y_l)$, $l = 1, \ldots, \mathcal{I}$. The other difficulty is the representation of the given function f. As pointed out earlier, we face the same issue in the application of HOSVD for high dimensions. The difficulty is not related to the application of the PGD algorithm for constructing the separated approximation $u^N(x_1, \ldots, x_D)$, but rather to the actual representation of the multidimensional function $f(x_1, \ldots, x_D)$.

3.4 The PGD for Data Post-Compression

As discussed later in this book, application of the PGD to models involving non-symmetric differential operators (such as parabolic and advection-diffusion equations) yields a separated representation of the solution that may involve too many terms. What we mean by too many is explained next.

For illustrative purposes, consider again a two-dimensional problem. Application of the PGD procedure gives the N-term approximation

$$u(x, y) = \sum_{i=1}^{N} X_i(x) \cdot Y_i(y), \tag{3.30}$$

with a residual error $\mathcal{E}(N)$.

Suppose now that the same problem has been solved by using any standard discretization technique, like finite differences or finite elements. This provides an approximate solution $u(x_j, y_i)$ at each grid or mesh point (x_j, y_i), from which we define the matrix $\mathbf{M}$ as in the previous sections. We could then compute the SVD of $\mathbf{M}$ and identify the number of terms $\tilde{N}$ in the SVD separated representation (3.2) of $\mathbf{M}$

$$\mathbf{M} \approx \sum_{i=1}^{\tilde{N}} \sigma_i \cdot \mathbf{U}_i \cdot \mathbf{V}_i^T, \tag{3.31}$$

such as to obtain a residual error identical to $\mathcal{E}(\mathcal{N})$.

Should N be found larger than $\tilde{N}$, then one would conclude that the PGD has computed too many terms and is thus not optimal for the problem under consideration.

For higher-dimensional problems, i.e. $D > 2$, a similar issue arises but the comparison is less clear since we know that the separated representation defined by the HOSVD is itself not necessarily optimal.

Having more terms than strictly needed is not necessarily an important drawback. Indeed, as we shall illustrate later in this book, the PGD is often applied only once and offline. In real-time applications, however, non-optimality can be detrimental. In this case, the natural question is: can a non-optimal separated representation be efficiently post-compressed by applying to it the PGD itself?

The answer is positive. For two-dimensional problems, the optimality of the PGD post-compression is indeed guaranteed in view of the equivalence between PGD-based separated approximation and SVD. For $D > 2$, the post-compressed solution can be computed efficiently but its optimality is not guaranteed.

Let us now detail the PGD post-compression procedure in the two-dimensional case. The idea is simply to apply the PGD approximation algorithm of the previous section to the non-optimal PGD solution

$$u(x, y) = \sum_{i=1}^{N} X_i(x) \cdot Y_i(y). \tag{3.32}$$

This solution is known, and it plays the role of the function $f(x, y)$ in the previous section.

The compressed representation $u^c(x, y)$ is then sought in the separated form

$$u^c(x, y) = \sum_{i=1}^{\tilde{N}} X_i^c(x) \cdot Y_i^c(y), \tag{3.33}$$

where we would expect that $\tilde{N}$ is significantly smaller than N.

For a particular enrichment step n, iteration p of the alternating direction scheme amounts to the following tasks:

- Calculating $X_n^{c,p}(x)$ from $Y_n^{c,p-1}(y)$, via the algebraic equation

$$\alpha^x \cdot X_n^{c,p}(x) = \sum_{j=1}^{N} \beta_j^x \cdot X_j(x) - \sum_{i=1}^{n-1} \gamma_i^x \cdot X_i^c(x), \tag{3.34}$$

where

$$\begin{cases} \alpha^x = \int_{\Omega_y} \left(Y_n^{c,p-1}(y) \right)^2 dy \\ \beta_j^x = \int_{\Omega_y} Y_n^{c,p-1}(y) \cdot Y_j(y) \, dy \\ \gamma_i^x = \int_{\Omega_y} Y_n^{c,p-1}(y) \cdot Y_i^c(y) \, dy \end{cases} . \tag{3.35}$$

- Calculating $Y_n^p(y)$ from the just-computed $X_n^p(x)$, via the algebraic equation

$$\alpha^y \cdot Y_n^{c,P}(y) = \sum_{j=1}^{N} \beta_j^y \cdot Y_i(y) - \sum_{i=1}^{n-1} \gamma_i^y \cdot Y_i^c(y), \qquad (3.36)$$

where

$$\begin{cases} \alpha^y = \int_{\Omega_x} \left(X_n^{c,P}(x) \right)^2 dx \\ \beta_j^y = \int_{\Omega_x} X_n^{c,P}(x) \cdot X_j(x) \, dx \\ \gamma_i^y = \int_{\Omega_x} X_n^{c,P}(x) \cdot X_i^c(x) \, dx \end{cases} . \qquad (3.37)$$

Since the function to be compressed is already in separated form (albeit probably sub-optimal), the two difficulties discussed at the end of the Sect. (i.e. representation of f and computation of the integrals β_j^x and β_j^y) do not arise at all in the present context.

Fig. 3.1 The authors of the present book (from *left* to *right*: AL, RK and FC). Original image and its separated approximation with 1, 15 and 50 terms

3.5 Numerical Example

We now illustrate the problem (3.6) of computing an approximation in separated form of a known function $f(x, y)$. In what follows, $f(x, y)$ is a function defined on a discrete 512 by 409 grid representing the pixel values of a grayscale image. In Fig. 3.1, we show the original image and its separated approximation with 1, 15 and 50 terms. One sees that even for complex functions the PGD (which is equivalent to the SVD in this case) can produce good quality separated approximations involving but few terms.

References

1. A. Ammar, M. Normandin, F. Daim, D. Gonzalez, E. Cueto, F. Chinesta, Non-incremental strategies based on separated representations: applications in computational rheology. Commun. Math. Sci. **8/3**, 671–695 (2010)
2. A. Ammar, F. Chinesta, A. Falco, On the convergence of a greedy rank-one update algorithm for a class of linear systems. Arch. Comput. Method. Eng. **17/4**, 473–486 (2010)
3. C. Le Bris, T. Lelièvre, Y. Maday, Results and questions on a nonlinear approximation approach for solving high-dimensional partial differential equations. Construct. Approx. **30**, 621–651 (2009)
4. T.G. Kolda, B.W. Bader, Tensor decompositions and applications. Technical Report SAND2007-6702, SANDIA National Laboratories (November 2007)

Chapter 4
The Transient Diffusion Equation

Abstract This chapter addresses the efficient solution of transient problems by considering space-time separated representations. In this case, the transient solution is calculated from a sequence of space and time problems. Such non-incremental solution procedure can lead to impressive computing-time savings, as discussed in detail. Finally, numerical examples are considered that demonstrate the efficiency of the proposed strategy.

Keywords Incremental time integration · Non-incremental time integration · Numerical stability · Proper Generalized Decomposition · Transient heat equation

We now go back to the case study of the diffusion equation, and extend it to time-dependent situations.

Much of the PGD procedure detailed in Chap. 2 immediately applies. We can thus be relatively concise and focus on the new issues that arise in this class of problems.

4.1 The One-Dimensional Transient Diffusion Equation

We start with the one-dimensional problem of computing the field $u(x, t)$ governed by

$$\frac{\partial u}{\partial t} - k \cdot \frac{\partial^2 u}{\partial x^2} = f, \tag{4.1}$$

in the space-time domain $\Omega = \Omega_x \times \Omega_t = (0, L) \times (0, \tau]$. The diffusivity k and source term f are assumed constant. We specify homogeneous initial and boundary conditions, i.e. $u(x, t = 0) = u(x = 0, t) = u(x = L, t) = 0$. More complex scenarios will be addressed later.

F. Chinesta et al., *The Proper Generalized Decomposition for Advanced Numerical Simulations*, SpringerBriefs in Applied Sciences and Technology, DOI: 10.1007/978-3-319-02865-1_4, © The Author(s) 2014

The weighted residual form of (4.1) reads

$$\int_{\Omega_x \times \Omega_t} u^* \cdot \left(\frac{\partial u}{\partial t} - k \cdot \frac{\partial^2 u}{\partial x^2} - f \right) dx \cdot dt = 0, \tag{4.2}$$

for all suitable test functions u^*.

Our objective is to obtain a PGD approximate solution in the separated form

$$u(x, t) = \sum_{i=1}^{N} X_i(x) \cdot T_i(t). \tag{4.3}$$

We do so by computing each term of the expansion at each step of an enrichment process, until a suitable stopping criterion is met.

4.1.1 Progressive Construction of the Separated Representation

At enrichment step n, the $n-1$ first terms of the PGD approximation (4.3) are known:

$$u^{n-1}(x, t) = \sum_{i=1}^{n-1} X_i(x) \cdot T_i(t). \tag{4.4}$$

We now wish to compute the next term $X_n(x) \cdot T_n(t)$ to get the enriched PGD solution

$$u^n(x, t) = u^{n-1}(x, t) + X_n(x) \cdot T_n(t) = \sum_{i=1}^{n-1} X_i(x) \cdot T_i(t) + X_n(x) \cdot T_n(t). \tag{4.5}$$

One must thus solve a non-linear problem for the unknown functions $X_n(x)$ and $T_n(t)$ by means of a suitable iterative scheme. We again rely on the simple but robust alternating direction scheme.

At enrichment step n, the PGD approximation $u^{n,p}$ obtained at iteration p is given by

$$u^{n,p}(x, t) = u^{n-1}(x, t) + X_n^p(x) \cdot T_n^p(t). \tag{4.6}$$

Starting from an arbitrary initial guess $T_n^0(t)$, the alternating direction strategy computes $X_n^p(x)$ from $T_n^{p-1}(t)$, and then $T_n^p(t)$ from $X_n^p(x)$. These non-linear iterations proceed until reaching a fixed point within a user-specified tolerance ϵ, i.e.

$$\| X_n^p(x) \cdot Y_n^p(y) - X_n^{p-1}(x) \cdot Y_n^{p-1}(y) \| < \epsilon, \tag{4.7}$$

where $\| \cdot \|$ is a suitable norm.

The enrichment step n thus ends with the assignments $X_n(x) \leftarrow X_n^p(x)$ and $T_n(t) \leftarrow T_n^p(t)$.

The enrichment process itself stops when an appropriate measure of error $\mathcal{E}(n)$ becomes small enough, i.e $\mathcal{E}(n) < \tilde{\epsilon}$. One can apply the stopping criteria discussed in Chap. 2.

Let us look at one particular alternating direction iteration at a given enrichment step.

4.1.2 Alternating Direction Strategy

Each iteration of the alternating direction scheme consists in the following two steps:

- Calculating $X_n^p(x)$ from $T_n^{p-1}(t)$

  At this stage, the approximation is given by

$$u^n(x, t) = \sum_{i=1}^{n-1} X_i(x) \cdot T_i(t) + X_n^p(x) \cdot T_n^{p-1}(t), \tag{4.8}$$

where all functions but $X_n^p(x)$ are known.

The simplest choice for the weight function u^* in (4.2) is

$$u^*(x, t) = X_n^*(x) \cdot T_n^{p-1}(t), \tag{4.9}$$

which amounts to consider a Galerkin formulation of the diffusion problem. Introducing (4.8) and (4.9) into (4.2), we obtain

$$\int_{\Omega_x \times \Omega_t} X_n^* \cdot T_n^{p-1} \cdot \left(X_n^p \cdot \frac{dT_n^{p-1}}{dt} - k \cdot \frac{d^2 X_n^p}{dx^2} \cdot T_n^{p-1} \right) dx \cdot dt$$

$$= - \int_{\Omega_x \times \Omega_t} X_n^* \cdot T_n^{p-1} \cdot \sum_{i=1}^{n-1} \left(X_i \cdot \frac{dT_i}{dt} - k \cdot \frac{d^2 X_i}{dx^2} \cdot T_i \right) dx \cdot dt$$

$$+ \int_{\Omega_x \times \Omega_t} X_n^* \cdot T_n^{p-1} \cdot f \, dx \cdot dt. \tag{4.10}$$

As all functions of time t are known, we can evaluate the following integrals:

$$\begin{cases} \alpha^x = \int_{\Omega_t} \left(T_n^{p-1}(t) \right)^2 dt \\[2mm] \beta^x = \int_{\Omega_t} T_n^{p-1}(t) \cdot \dfrac{d T_n^{p-1}(t)}{dt} \, dt \\[2mm] \gamma_i^x = \int_{\Omega_t} T_n^{p-1}(t) \cdot T_i(t) \, dt \\[2mm] \delta_i^x = \int_{\Omega_t} T_n^{p-1}(t) \cdot \dfrac{d T_i(t)}{dt} \, dt \\[2mm] \xi^x = \int_{\Omega_t} T_n^{p-1}(t) \cdot f \, dt \end{cases} \qquad (4.11)$$

Equation (4.10) then takes the form

$$\int_{\Omega_x} X_n^* \cdot \left(-k \cdot \alpha^x \cdot \frac{d^2 X_n^p}{dx^2} + \beta^x \cdot X_n^p \right) dx$$

$$= \int_{\Omega_x} X_n^* \cdot \sum_{i=1}^{n-1} \left(k \cdot \gamma_i^x \cdot \frac{d^2 X_i}{dx^2} - \delta_i^x \cdot X_i \right) dx + \int_{\Omega_x} X_n^* \cdot \xi^x \, dx. \qquad (4.12)$$

This defines a one-dimensional boundary value problem (BVP), which is readily solved by means of a standard finite element method to obtain an approximation of the function X_n^p. As another option, one can go back to the associated strong form

$$-k \cdot \alpha^x \cdot \frac{d^2 X_n^p}{dx^2} + \beta^x \cdot X_n^p = \sum_{i=1}^{n-1} \left(k \cdot \gamma_i^x \cdot \frac{d^2 X_i}{dx^2} - \delta_i^x \cdot X_i \right) + \xi^x, \qquad (4.13)$$

and then solve it using any suitable numerical method, such as finite differences for example. The strong form (4.13) is a second-order differential equation for X_n^p due to the fact that the original diffusion equation (4.1) involves a second-order x-derivative of the unknown field u.

The homogeneous Dirichlet boundary conditions $X_n^p(x = 0) = X_n^p(x = L) = 0$ are readily specified with either weak or strong formulations.

- Calculating $T_n^p(t)$ from the just-computed $X_n^p(x)$

The procedure mirrors what we have just done. It suffices to exchange the roles played by the relevant functions of x and t.

The current PGD approximation reads

$$u^n(x, t) = \sum_{i=1}^{n-1} X_i(x) \cdot T_i(t) + X_n^p(x) \cdot T_n^p(t), \qquad (4.14)$$

where all functions are known except $T_n^p(t)$.

With the Galerkin weight function

$$u^*(x, t) = X_n^p(x) \cdot T_n^*(t), \qquad (4.15)$$

the weighted residual form (4.2) becomes

$$
\int_{\Omega_x \times \Omega_t} X_n^p \cdot T_n^* \cdot \left(X_n^p \cdot \frac{dT_n^p}{dt} - k \cdot \frac{d^2 X_n^p}{dx^2} \cdot T_n^p \right) dx \cdot dt
$$

$$
= - \int_{\Omega_x \times \Omega_t} X_n^p \cdot T_n^* \cdot \sum_{i=1}^{n-1} \left(X_i \cdot \frac{dT_i}{dt} - k \cdot \frac{d^2 X_i}{dx^2} \cdot T_i \right) dx \cdot dt
$$

$$
+ \int_{\Omega_x \times \Omega_t} X_n^p \cdot T_n^* \cdot f \, dx \cdot dt. \tag{4.16}
$$

Since all functions of x are known, we can perform the following integrals

$$
\begin{cases}
\alpha^t = \int_{\Omega_x} \left(X_n^p(x) \right)^2 dx \\
\beta^t = \int_{\Omega_x} X_n^p(x) \cdot \dfrac{d^2 X_n^p(x)}{dx^2} dx \\
\gamma_i^t = \int_{\Omega_x} X_n^p(x) \cdot X_i(x) \, dx \\
\delta_i^t = \int_{\Omega_x} X_n^p(x) \cdot \dfrac{d^2 X_i(x)}{dx^2} dx \\
\xi^t = \int_{\Omega_x} X_n^p(x) \cdot f \, dx
\end{cases} \tag{4.17}
$$

Equation (4.16) then becomes

$$
\int_{\Omega_t} T_n^* \cdot \left(\alpha^t \cdot \frac{dT_n^p}{dt} - k \cdot \beta^t \cdot T_n^p \right) dt
$$

$$
= \int_{\Omega_t} T_n^* \cdot \sum_{i=1}^{n-1} \left(-\gamma_i^t \cdot \frac{dT_i}{dt} + k \cdot \delta_i^t \cdot T_i \right) dt + \int_{\Omega_t} T_n^* \cdot \xi^t \, dt. \tag{4.18}
$$

We have thus obtained an initial value problem (IVP) for the function T_n^p. The weighted residual form (4.18) can be solved by means of any stabilized finite element scheme (e.g discontinuous Galerkin). The associated strong form reads

$$
\alpha^t \cdot \frac{dT_n^p}{dt} - k \cdot \beta^t \cdot T_n^p = \sum_{i=1}^{n-1} \left(-\gamma_i^t \cdot \frac{dT_i}{dt} + k \cdot \delta_i^t \cdot T_i \right) + \xi^t. \tag{4.19}
$$

Since the original diffusion equation involves a first-order derivative of u with respect to t, we have thus obtained a first-order ordinary differential equation for T_n^p. Any classical numerical technique can be used to solve it.

The initial condition $T_n^p(t = 0) = 0$ is readily specified with either weak or strong form.

4.1.3 Non-Incremental Versus Incremental Time Integrations

It is useful to reflect on the considerable difference between the above PGD strategy and traditional, incremental time integration schemes.

Indeed, the PGD allows for a *non-incremental* solution of time-dependent problems. Let $\mathcal{Q}_n$ denote the number of non-linear iterations of the alternating direction algorithm required to compute the new term $X_n(x) \cdot T_n(t)$ at enrichment step n. Then, the entire PGD procedure to obtain the N-term approximation (4.3) involves the solution of a total of $\mathcal{Q} = (\mathcal{Q}_1 + \cdots + \mathcal{Q}_N)$ *decoupled*, one-dimensional boundary and initial value problems. The BVP's are defined over the space domain Ω_x, and their computational complexity scales with the size of the one-dimensional mesh used to discretize them. The IVP's are defined over the time interval Ω_t, and their complexity is usually negligible compared to that of the BVP's, even when extremely small time steps are used for their discretization.

This is vastly different from a standard, incremental solution procedure. If P is the total number of time steps for the complete simulation, i.e. $P = \tau/\Delta t$, an incremental procedure involves the solution of a BVP in Ω_x at each time step, i.e. a total of P BVP's. This can be a very large number indeed, as the time step Δt must be chosen small enough to guarantee the stability of the numerical scheme.

Numerical experiments with the PGD show that the $\mathcal{Q}_n$'s rarely exceed ten, while N is a few tens. Thus, the complexity of the complete PGD solution is a few hundreds of BVP solutions in Ω_x. This is generally several orders of magnitude less than the total of P BVP's that must be solved using a standard incremental procedure.

4.2 Multi-Dimensional Transient Diffusion Equation

We now consider the multi-dimensional version of the transient diffusion equation,

$$\frac{\partial u}{\partial t} - k \cdot \Delta u = f(\mathbf{x}, t), \qquad (4.20)$$

defined in the space-time domain $\Omega = \Omega_\mathbf{x} \times \Omega_t = \Omega_\mathbf{x} \times (0, \tau]$. We consider three-dimensional spatial domains $\Omega_\mathbf{x} \subset \mathcal{R}^3$, with boundary Γ. As in the previous section, the diffusivity k and source term f are assumed constant, and homogeneous initial and boundary conditions are specified, i.e. $u(\mathbf{x}, t = 0) = u(\mathbf{x} \in \Gamma, t) = 0$.

The starting point of the PGD strategy is once again the weighted residual formulation of the problem,

$$\int_{\Omega_x \times \Omega_t} u^* \cdot \left(\frac{\partial u}{\partial t} - k \cdot \Delta u - f \right) dx \cdot dt = 0. \qquad (4.21)$$

Now, three possibilities can be envisaged:

1. The spatial domain Ω_x is non separable.
   In this case, the space coordinates $\mathbf{x} = (x, y, z)$ cannot be separated, and the PGD solution is sought in the form

$$u(\mathbf{x}, t) = \sum_{i=1}^{N} X_i(\mathbf{x}) \cdot T_i(t). \tag{4.22}$$

The procedure of the previous section is then readily emulated by considering $\mathbf{x}$ instead of x and Δu instead of $\frac{\partial^2 u}{\partial x^2}$. The reader will thus easily verify that iteration p of the alternating direction strategy at a given enrichment step n consists in the following two tasks.
First, one must solve in Ω_x the following three-dimensional BVP to obtain the unknown function X_n^p,

$$\alpha^x \cdot \Delta X_n^p + \beta^x \cdot X_n^p = f_n^p. \tag{4.23}$$

Here, α^x, β^x and $f_n^p(\mathbf{x})$ are integrals over Ω_t of combinations of the source term f and the known functions T_n^{p-1} and T_i for $i < n$.
The second task yields the function $T_n^p(t)$. One must solve within Ω_t an IVP of the form

$$\alpha^t \cdot \frac{dT_n^p}{dt} + \beta^t \cdot T_n^p = g_n^p, \tag{4.24}$$

where α^t, β^t and $g_n^p(t)$ are integrals over Ω_x of combinations of the source term f and the known functions X_n^p and X_i for $i < n$.
It is again important to emphasize the considerable difference between the PGD non-incremental strategy and traditional, incremental time integration schemes. If $\mathcal{Q}_n$ denotes the number of non-linear iterations of the alternating direction algorithm required to compute the new term $X_n(\mathbf{x}) \cdot T_n(t)$ at enrichment step n, then the entire PGD procedure to obtain the N-term approximation (4.22) involves the solution of a total of $\mathcal{Q} = (\mathcal{Q}_1 + \cdots + \mathcal{Q}_N)$ decoupled, three-dimensional BVP's and one-dimensional IVP's. The BVP's are defined over the space domain Ω_x, and their computational complexity scales with the size of the three-dimensional mesh used to discretize them. The IVP's are defined over the time interval Ω_t, and their complexity is negligible compared to that of the BVP's, even for extremely small time steps.
Numerical experiments with the PGD show that the $\mathcal{Q}_n$'s rarely exceed ten, while N is a few tens. Thus, the complexity of the complete PGD solution is a few hundreds of 3D BVP solutions in Ω_x. A standard, incremental solution procedure with a total for P time steps involves a total of P three-dimensional BVP's, which is generally several orders of magnitude more than for the PGD strategy.

2. The spatial domain $\Omega_{\mathbf{x}}$ is partially separable

By this we mean that two dimensions can be separated from the third-one. Assume for example that $\Omega_{\mathbf{x}} = \Omega_{xy} \times \Omega_z$, with $\Omega_{xy} \subset \mathcal{R}^2$ and $\Omega_z \subset \mathcal{R}$. The PGD solution is then thought in the form

$$u(\mathbf{x}, z, t) = \sum_{i=1}^{N} X_i(\mathbf{x}) \cdot Z_i(z) \cdot T_i(t), \qquad (4.25)$$

where $\mathbf{x} = (x, y) \in \Omega_{xy}$ and $z \in \Omega_z$.

The reader will thus easily verify that iteration p of the alternating direction strategy at a given enrichment step n consists in the following three tasks:

First, solve in Ω_{xy} the following two-dimensional BVP to obtain the function X_n^p,

$$\alpha^x \cdot \Delta X_n^p + \beta^x \cdot X_n^p = f_n^p, \qquad (4.26)$$

where α^x, β^x and $f_n^p(\mathbf{x})$ are integrals over $\Omega_z \times \Omega_t$ of combinations of the source term f and the known functions Z_n^{p-1}, T_n^{p-1}, Z_i and T_i for $i < n$.

Second, solve in Ω_z the following one-dimensional BVP to obtain the function Z_n^p,

$$\alpha^z \cdot \frac{d^2 Z_n^p}{dz^2} + \beta^z \cdot Z_n^p = h_n^p, \qquad (4.27)$$

where α^z, β^z and $h_n^p(z)$ are integrals over $\Omega_{xy} \times \Omega_t$ of combinations of the source term f and the known functions X_n^p, T_n^{p-1}, X_i and T_i for $i < n$.

Third, solve in Ω_t the following one-dimensional IVP to obtain the function T_n^p,

$$\alpha^t \cdot \frac{dT_i^p}{dt} + \beta^t \cdot T_n^p = g_n^p, \qquad (4.28)$$

where α^t, β^t and $g_n^p(t)$ are integrals over $\Omega_{xy} \times \Omega_z$ of combinations of the source term f and the known functions X_n^p, Z_n^p, X_i and Z_i for $i < n$.

We can repeat our discussion regarding the complexity of this PGD non-incremental strategy versus standard incremental schemes. Clearly, what will dominate the cost of the PGD procedure is the total of $\mathcal{Q}$ two-dimensional BVP's to be solved in Ω_{xy}. The BVP's in Ω_z and IVP's in Ω_t being one-dimensional, their complexity is comparatively negligible. Thus, the computational cost of the PGD simulation will be orders of magnitude smaller than that of a standard incremental procedure, which requires the solution of a three-dimensional BVP at each time step.

The cost savings provided by the PGD are even higher when the spatial domain is fully separable, as we now consider.

3. The spatial domain $\Omega_{\mathbf{x}}$ is fully separable

We now have $\Omega_{\mathbf{x}} = \Omega_x \times \Omega_y \times \Omega_z$, and the PGD solution takes the fully-separated form

$$u(x, y, z, t) = \sum_{i=1}^{N} X_i(x) \cdot Y_i(y) \cdot Z_i(z) \cdot T_i(t). \tag{4.29}$$

Iteration p of the alternating direction strategy at a given enrichment step n consists in the following four tasks:

First, solve in Ω_x the following one-dimensional BVP to obtain the function X_n^p,

$$\alpha^x \cdot \frac{d^2 X_n^p}{dx^2} + \beta^x \cdot X_n^p = f_n^p, \tag{4.30}$$

where α^x, β^x and $f_n^p(x)$ involve integrals over $\Omega_y \times \Omega_z \times \Omega_t$ of combinations of the source term f and the known functions Y_n^{p-1}, Z_n^{p-1}, T_n^{p-1}, Y_i, Z_i and T_i for $i < n$.

Second, solve in Ω_y the following one-dimensional BVP to obtain the function Y_n^p,

$$\alpha^y \cdot \frac{d^2 Y_n^p}{dy^2} + \beta^y \cdot Y_n^p = w_n^p, \tag{4.31}$$

where α^y, β^y and $w_n^p(y)$ involve integrals over $\Omega_x \times \Omega_z \times \Omega_t$ of combinations of the source term f and the known functions X_n^p, Z_n^{p-1}, T_n^{p-1}, X_i, Z_i and T_i for $i < n$.

Third, solve in Ω_z the following one-dimensional BVP to obtain the function Z_n^p,

$$\alpha^z \cdot \frac{d^2 Z_n^p}{dz^2} + \beta^z \cdot Z_n^p = h_n^p, \tag{4.32}$$

where α^z, β^z and $h_n^p(z)$ involve integrals over $\Omega_x \times \Omega_y \times \Omega_t$ of combinations of the source term f and the known functions X_n^p, Y_n^p, T_n^{p-1}, X_i, Y_i and T_i for $i < n$.

Fourth, solve in Ω_t the following one-dimensional IVP to obtain the function T_n^p,

$$\alpha^t \cdot \frac{dT_i^p}{dt} + \beta^t \cdot T_n^p = g_n^p, \tag{4.33}$$

where α^t, β^t and $g_n^p(t)$ involve integrals over $\Omega_x \times \Omega_y \times \Omega_z$ of combinations of the source term f and the known functions X_n^p, Y_n^p, Z_n^p, X_i, Y_i and Z_i for $i < n$.

The cost savings provided by the PGD are potentially phenomenal when the spatial domain is fully separable. Indeed, the complexity of the PGD simulation now scales with the *one-dimensional* meshes used to solve the BVP's in Ω_x, Ω_y and Ω_z, regardless of the time step used in the solution of the decoupled IVP's in Ω_t. The computational cost is thus orders of magnitude smaller than that of a standard incremental procedure, which requires the solution of a three-dimensional BVP at each time step.

4.3 Numerical Example

Let us consider a particular case of the transient diffusion equation (4.1):

$$\frac{\partial u}{\partial t} - k \cdot \frac{\partial^2 u}{\partial x^2} = f. \tag{4.34}$$

The problem is defined in the space-time domain $\Omega = \Omega_x \times \Omega_t = (0, 1) \times (0, 0.1]$. For the diffusivity and the source term, we take $k = 1$ and $f = 1$ respectively. We specify homogeneous initial and boundary conditions, i.e. $u(x, t = 0) = u(x = 0, t) = u(x = 1, t) = 0$. The corresponding exact solution can be found analytically:

$$u_{ex}(x, t) = \frac{x \cdot (1 - x)}{2} - \sum_{n \text{ odd}} \frac{4}{\pi^3 \cdot n^3} \cdot \sin(n \cdot \pi \cdot x) \cdot e^{-n^2 \cdot \pi^2 \cdot t}. \tag{4.35}$$

The unknown functions $X_i(x)$ and $T_i(t)$ are sought on uniform grids with 61 points for the space domain and 151 points for the time domain. All spatial one-dimensional BVP's arising in the solution procedure are solved using second-order finite differences. The IVP's are solved using the Backward Euler method. All integrals are evaluated numerically by means of the trapezoidal rule.

In Figs. 4.1 and 4.2, we show the normalized functions $X_i(x)$ and $T_i(t)$ for $i = 1, \ldots, 4$. Again, one can observe that, as i increases, both $X_i(x)$ and $T_i(t)$ account for a higher frequency content of the numerical solution.

Fig. 4.1 Normalized functions $X_i(x)$ for $i = 1, \ldots, 4$ produced by the PGD solution of (4.1)

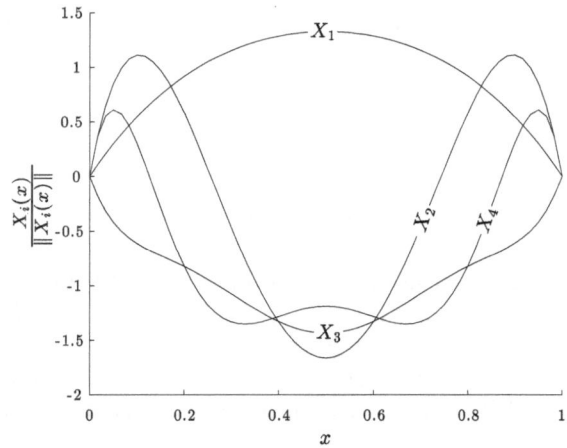

Fig. 4.2 Normalized functions $T_i(t)$ for $i = 1, \ldots, 4$ produced by the PGD solution of (4.1)

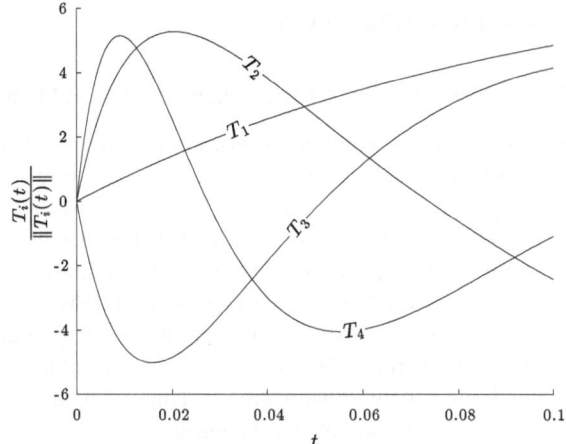

To illustrate the convergence of the PGD solution to the analytical solution, we define the following quadratic error between the analytical solution $u_{ex}(x, t)$ and a numerical PGD solution $u^N(x, t)$ with N enrichment steps:

$$E(u^N) = \int_0^{\tilde{0.1}} \int_0^{\tilde{1}} \left(u_{ex}(x, t) - u^N(x, t) \right)^2 dx \cdot dt. \tag{4.36}$$

Here, the symbol $\tilde{\int}$ refers to a numerical integration carried out with the trapezoidal rule on the nodal values. In Fig. 4.3, we observe that the global convergence of the PGD solution towards the analytical solution is no longer monotonous.

Fig. 4.3 PGD error $E(u^N)$ as a function of the number N of enrichment steps

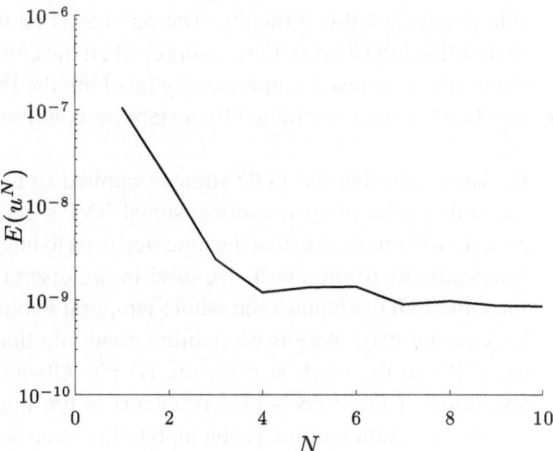

4.4 Concluding Remarks

We conclude this chapter with a number of useful remarks.

- Accounting for non-homogenous initial and/or boundary conditions
  Non-homogeneous initial or boundary conditions are handled as described in Chap. 2. It suffices to define a function verifying these conditions, compute a separated representation of it, and then consider this representation as the first modes of the PGD approximation of the unknown solution.
- Accounting for Neumann boundary conditions
  As described in Chap. 2, this type of boundary conditions is treated via an integration by parts of the model's weighted residual form. The Neumann boundary condition is then implemented as a natural boundary condition.
- Optimal separation of space coordinates
  One should not conclude from the discussion of the previous section that the complete separation of the spatial domain is necessarily optimal in terms of computational cost. Indeed, for a given level of accuracy, the number of terms N in the PGD expansion generally increases with the number of coordinates that are separated in the solution representation. For example, a function of two coordinates can be represented by a single function, i.e. itself, but when it is separated as a sum of products of one-dimensional functions, many terms are usually needed. The optimal level of separation of the spatial domain will depend on the considered problem and on the expected use of the computed PGD solution.
- Non-symmetry of the differential operator
  The parabolic differential operator of the transient diffusion equation is non symmetric, as it contains a first-order time derivative. The PGD is not expected to be optimal when non-symmetry dominates, i.e. for problems with a small diffusivity k. Indeed, the number of terms involved in the separated representation will increase inordinately in the limit of very small diffusivity. Fortunately, alternative PGD strategies based on residual minimization or the one proposed in [1] are available to alleviate this difficulty. The one based on the residual minimization will be described in Chap. 6. Furthermore, when the computed solution contains many terms, it can be post-compressed by invoking the PGD, as indicated in Chap. 2.
- The PGD is not constrained by a stability condition linking time step and mesh size
  We have seen that the PGD strategy applied to transient problems involves the decoupled solution of low-dimensional BVP's and one-dimensional IVP's. It is important to emphasize that the time step used to integrate the IVP's can be selected *independently* of the mesh size used in the discretization of the BVP's. Indeed, since the PGD computes the whole temporal evolution of the solution in a non-incremental way, there is no stability condition that would link the time step for the IVP's to the mesh size for the BVP's. Obviously, if the space and/or time resolution of these decoupled problems is too poor, the PGD will compute an inaccurate solution of the problem, but this issue regards convergence rather than stability.

- The PGD is a flexible numerical framework

  In terms of numerical software engineering, the PGD can be viewed as an *outer shell* which organizes in a rather flexible manner a series of solvers for decoupled BVP's and IVP's. The user can indeed select, and even re-use as such, whatever preferred solver is available for a particular BVP or IVP. For example, finite differences could be used for the IVP's, whereas finite elements or finite volumes could be considered for solving the BVP's. In fact, any combination of suitable numerical approaches can be implemented within a PGD framework.

Reference

1. M. Billaud-Friess, A. Nouy, O. Zahm, A tensor approximation method based on ideal minimal residual formulations for the solution of high dimensional problems. arXiv:1304.6126 (in press)

Chapter 5
Parametric Models

Abstract Since separated representations allow one to circumvent the curse of dimensionality, one can consider model parameters, boundary conditions, initial conditions or geometrical parameters defining the computational domain, as extra-coordinates of the problem. Thus, standard models become multi-dimensional, but by solving them only once and offline using the PGD, the solution of the model is available for any choice of the parameters considered as extra-coordinates. This parametric solution can then be used online for different purposes, such as real time simulation, efficient optimization or inverse analysis, or simulation-based control. In this chapter, we illustrate the procedures for considering (a) model parameters, (b) constant and non-constant Dirichlet and Neumann boundary conditions, (c) initial conditions and (d) geometrical parameters, as extra-coordinates of a resulting multi-dimensional model.

Keywords Geometrical parameters · Material parameters · Parametric boundary conditions · Parametric model · Parametric solution · Proper Generalized Decomposition

In this chapter, we show in detail how computational models can easily be enriched in the PGD framework by introducing several of the problem parameters as extra-coordinates. The PGD thus produces a *general solution* that can be post-processed at will for various purposes, such as optimization or inverse identification.

5.1 Material Parameters as Extra-Coordinates

First, we discuss the possibility of introducing material parameters as extra-coordinates. As in the previous chapter, we consider the transient diffusion equation

$$\frac{\partial u}{\partial t} - k \cdot \Delta u = f, \tag{5.1}$$

F. Chinesta et al., *The Proper Generalized Decomposition for Advanced Numerical Simulations*, SpringerBriefs in Applied Sciences and Technology,
DOI: 10.1007/978-3-319-02865-1_5, © The Author(s) 2014

with uniform diffusivity k, uniform source term f, and homogeneous initial and Dirichlet boundary conditions. We wish to compute *at once* a general solution of the problem for *all* values of k in a given interval of values Ω_k.

In the PGD framework, we thus consider the diffusivity k as an extra-coordinate of the problem, in addition to space $\mathbf{x}$ and time t. The problem (5.1) is now defined for $(\mathbf{x}, t, k) \in \Omega \times \Omega_t \times \Omega_k$, with $\Omega \subset \mathcal{R}^3$, $\Omega_t \subset \mathcal{R}$ and $\Omega_k \subset \mathcal{R}$.

Thus, instead of solving a series of diffusion problems for different discrete values of the diffusivity parameter, we wish to solve at once a more general problem. The price to pay is of course an increase of the problem dimensionality. This is not a major issue for the PGD, whose computational complexity scales only linearly (and not exponentially) with the space dimension.

It is rather straightforward to take account of the extra-coordinate in the PGD framework. We start from the weighted residual form of (5.1), which reads

$$
\int_{\Omega \times \Omega_t \times \Omega_k} u^* \cdot \left(\frac{\partial u}{\partial t} - k \cdot \Delta u - f \right) dx \cdot dt \cdot dk = 0, \tag{5.2}
$$

for all suitable test functions u^*.

The PGD solution is then sought in the separated form

$$
u(\mathbf{x}, t, k) = \sum_{i=1}^{N} X_i(\mathbf{x}) \cdot T_i(t) \cdot K_i(k), \tag{5.3}
$$

where we have left the space coordinates $\mathbf{x}$ un-separated in this particular illustration. Each term of the expansion is computed one at a time, thus enriching the PGD solution until a suitable convergence criterion is satisfied

At enrichment step n of the PGD algorithm, the following approximation is already known,

$$
u^{n-1}(\mathbf{x}, t, k) = \sum_{i=1}^{n-1} X_i(\mathbf{x}) \cdot T_i(t) \cdot K_i(k). \tag{5.4}
$$

We wish to compute the next functional product $X_n(\mathbf{x}) \cdot T_n(t) \cdot K_n(k)$, which we write as $R(\mathbf{x}) \cdot S(t) \cdot W(k)$ for notational simplicity.

Thus, the solution at enrichment step n reads

$$
u^n = u^{n-1} + R(\mathbf{x}) \cdot S(t) \cdot W(k). \tag{5.5}
$$

The weighted residual form (5.2) yields a non-linear problem for the unknown functions R, S and W, which we solve iteratively by means of an alternating direction scheme. Each iteration consists of three steps that are repeated until a fixed point is reached.

The first step assumes $S(t)$ and $W(k)$ known from the previous iteration and computes an update for $R(\mathbf{x})$. To do so, the weighted residual form can be integrated in $\Omega_t \times \Omega_k$ since all functions of time t and diffusivity k are assumed known at the present step. The evaluation of the integral is computationally very cheap as it involves separated representations. Indeed, consider the integral over $\Omega_t \times \Omega_k$ of a generic function $F(\mathbf{x}, t, k)$,

$$\int_{\Omega_t \times \Omega_k} F(\mathbf{x}, t, k) \, dt \cdot dk. \tag{5.6}$$

In the general case, one should compute an integral in $\Omega_t \times \Omega_k$ for each value of $\mathbf{x}$. Even when considering only a discrete number of points $\mathbf{x}_k$, the integration complexity scales with the number of points $\mathbf{x}_k$. When the integrand is expressed in separated form, however, the integral reduces to

$$\int_{\Omega_t \times \Omega_k} F(\mathbf{x}, t, k) \, dt \cdot dk = \int_{\Omega_t \times \Omega_k} \sum_{j=1}^{\mathcal{F}} F_j^x(\mathbf{x}) \cdot F_j^t(t) \cdot F_j^k(k) \, dt \cdot dk$$

$$= \sum_{j=1}^{\mathcal{F}} F_j^x(\mathbf{x}) \cdot \left(\int_{\Omega_t} F_j^t(t) \, dt \right) \cdot \left(\int_{\Omega_k} F_j^k(k) \, dk \right). \tag{5.7}$$

The task is thus reduced to a number $2\mathcal{F}$ of one-dimensional integrals.

The other two steps of a particular non-linear iteration proceed in the usual manner. From the just-updated $R(\mathbf{x})$ and the previously-used $W(k)$, we update $S(t)$. Finally, from the just-updated $R(\mathbf{x})$ and $S(t)$, we update $W(k)$. Here again, the separability of the functions to be integrated is a key point from the computational point of view.

This iterative procedure continues until reaching convergence. The converged functions R, S and W yield the new functional product at the current enrichment step, i.e $X_n(\mathbf{x}) \leftarrow R(\mathbf{x})$, $T_n(t) \leftarrow S(t)$, and $K_n(k) \leftarrow W(k)$.

The explicit form of these operations is described in what follows.

- Computing $R(\mathbf{x})$ from $S(t)$ and $W(k)$
  In the weighted residual form (5.2), the trial and test functions are respectively given by

$$u^n(\mathbf{x}, t, k) = \sum_{i=1}^{n-1} X_i(\mathbf{x}) \cdot T_i(t) \cdot K_i(k) + R(\mathbf{x}) \cdot S(t) \cdot W(k), \tag{5.8}$$

and

$$u^*(\mathbf{x}, t, k) = R^*(\mathbf{x}) \cdot S(t) \cdot W(k), \tag{5.9}$$

wherein the functions S and W are known from the previous iteration. Introducing (5.8) and (5.9) into (5.2) yields

$$\int\limits_{\Omega\times\Omega_t\times\Omega_k} R^* \cdot S \cdot W \cdot \left(R \cdot \frac{\partial S}{\partial t} \cdot W - k \cdot \Delta R \cdot S \cdot W \right) \, d\mathbf{x} \cdot dt \cdot dk$$
$$= - \int\limits_{\Omega\times\Omega_t\times\Omega_k} R^* \cdot S \cdot W \cdot \mathcal{R}^{n-1} \, d\mathbf{x} \cdot dt \cdot dk, \tag{5.10}$$

where $\mathcal{R}^{n-1}$ is the residual related to $u^{n-1}(\mathbf{x}, t, k)$,

$$\mathcal{R}^{n-1} = \sum_{i=1}^{n-1} X_i \cdot \frac{\partial T_i}{\partial t} \cdot K_i - \sum_{i=1}^{n-1} k \cdot \Delta X_i \cdot T_i \cdot K_i - f. \tag{5.11}$$

Since all functions of time t and diffusivity k are known, we can integrate (5.10) over $\Omega_t \times \Omega_k$. Defining the following notations,

$$\begin{bmatrix} w_1 = \int\limits_{\Omega_k} W^2 dk & s_1 = \int\limits_{\Omega_t} S^2 dt & r_1 = \int\limits_{\Omega} R^2 d\mathbf{x} \\ w_2 = \int\limits_{\Omega_k} kW^2 dk & s_2 = \int\limits_{\Omega_t} S \cdot \frac{dS}{dt} dt & r_2 = \int\limits_{\Omega} R \cdot \Delta R \, d\mathbf{x} \\ w_3 = \int\limits_{\Omega_k} W \, dk & s_3 = \int\limits_{\Omega_t} S \, dt & r_3 = \int\limits_{\Omega} R \, d\mathbf{x} \\ w_4^i = \int\limits_{\Omega_k} W \cdot K_i \, dk & s_4^i = \int\limits_{\Omega_t} S \cdot \frac{dT_i}{dt} dt & r_4^i = \int\limits_{\Omega} R \cdot \Delta X_i \, d\mathbf{x} \\ w_5^i = \int\limits_{\Omega_k} kW \cdot K_i \, dk & s_5^i = \int\limits_{\Omega_t} S \cdot T_i \, dt & r_5^i = \int\limits_{\Omega} R \cdot X_i \, d\mathbf{x} \end{bmatrix}, \tag{5.12}$$

Equation (5.10) becomes

$$\int\limits_{\Omega} R^* \cdot (w_1 \cdot s_2 \cdot R - w_2 \cdot s_1 \cdot \Delta R) \, d\mathbf{x}$$
$$= - \int\limits_{\Omega} R^* \cdot \left(\sum_{i=1}^{n-1} w_4^i \cdot s_4^i \cdot X_i - \sum_{i=1}^{n-1} w_5^i \cdot s_5^i \cdot \Delta X_i - w_3 \cdot s_3 \cdot f \right) \, d\mathbf{x}. \tag{5.13}$$

Equation (5.27) is the weighted residual form of an elliptic steady-state BVP for the unknown function $R(\mathbf{x})$. We can solve it by means of any suitable discretization technique such as finite elements or finite volumes. Another possibility is to consider the corresponding strong form,

$$w_1 \cdot s_2 \cdot R - w_2 \cdot s_1 \cdot \Delta R$$

$$= - \left(\sum_{i=1}^{n-1} w_4^i \cdot s_4^i \cdot X_i - \sum_{i=1}^{n-1} w_5^i \cdot s_5^i \cdot \Delta X_i - w_3 \cdot s_3 \cdot f \right). \tag{5.14}$$

There is again a wide choice of techniques, such as finite differences, to solve this equation numerically.

- Computing $S(t)$ from $R(\mathbf{x})$ and $W(k)$

With the test function now given by

$$u^*(\mathbf{x}, t, k) = S^*(t) \cdot R(\mathbf{x}) \cdot W(k), \tag{5.15}$$

the weighted residual form (5.2) reads

$$\int_{\Omega \times \Omega_t \times \Omega_k} S^* \cdot R \cdot W \cdot \left(R \cdot \frac{\partial S}{\partial t} \cdot W - k \cdot \Delta R \cdot S \cdot W \right) dx \cdot dt \cdot dk$$
$$= - \int_{\Omega \times \Omega_t \times \Omega_k} S^* \cdot R \cdot W \cdot \mathcal{R}^{n-1} \, dx \cdot dt \cdot dk. \tag{5.16}$$

By integrating over $\Omega \times \Omega_k$ and using the notations (5.12), we obtain

$$\int_{\Omega_t} S^* \cdot \left(w_1 \cdot r_1 \cdot \frac{dS}{dt} - w_2 \cdot r_2 \cdot S \right) dt$$
$$= - \int_{\Omega_t} S^* \cdot \left(\sum_{i=1}^{n-1} w_4^i \cdot r_5^i \cdot \frac{dT_i}{dt} - \sum_{i=1}^{n-1} w_5^i \cdot r_4^i \cdot T_i - w_3 \cdot r_3 \cdot f \right) dt. \tag{5.17}$$

This is the weighted residual form of an IVP for the unknown function $S(t)$, that can be solved by means of any stabilized finite element technique (e.g. Streamline-Upwind or Discontinuous Galerkin). The corresponding strong form reads

$$w_1 \cdot r_1 \cdot \frac{dS}{dt} - w_2 \cdot r_2 \cdot S$$
$$= - \left(\sum_{i=1}^{n-1} w_4^i \cdot r_5^i \cdot \frac{dT_i}{dt} - \sum_{i=1}^{n-1} w_5^i \cdot r_4^i \cdot T_i - w_3 \cdot r_3 \cdot f \right), \tag{5.18}$$

and it can be solved by means of any suitable technique for ordinary differential equations (e.g. backward finite differences, or higher-order Runge-Kutta schemes).

- Computing $W(k)$ from $R(\mathbf{x})$ and $S(t)$

In this third and final step, the test function is given by

$$u^*(\mathbf{x}, t, k) = W^*(k) \cdot R(\mathbf{x}) \cdot S(t), \tag{5.19}$$

and the weighted residual form (5.2) becomes

$$\int_{\Omega \times \Omega_t \times \Omega_k} W^* \cdot R \cdot S \cdot \left(R \cdot \frac{\partial S}{\partial t} \cdot W - k \cdot \Delta R \cdot S \cdot W \right) dx \cdot dt \cdot dk$$
$$= - \int_{\Omega \times \Omega_t \times \Omega_k} W^* \cdot R \cdot S \cdot \mathcal{R}^{n-1} \, dx \cdot dt \cdot dk. \tag{5.20}$$

Integration over $\Omega \times \Omega_t$ and use of the notations (5.12) yield

$$
\int\limits_{\Omega_k} W^* \cdot (r_1 \cdot s_2 \cdot W - r_2 \cdot s_1 \cdot k \cdot W) \; dk
$$

$$
= - \int\limits_{\Omega_k} W^* \cdot \left(\sum_{i=1}^{n-1} r_5^i \cdot s_4^i \cdot K_i - \sum_{i=1}^{n-1} r_4^i \cdot s_5^i \cdot k \cdot K_i - r_3 \cdot s_3 \cdot f \right) dk. \tag{5.21}
$$

Not surprisingly, this equation does not involve any differential operator since the original model (5.1) does not contain derivatives with respect to the diffusivity. The corresponding strong form reads

$$
(r_1 \cdot s_2 - r_2 \cdot s_1 \cdot k) \cdot W = - \left(\sum_{i=1}^{n-1} \left(r_5^i \cdot s_4^i - r_4^i \cdot s_5^i \cdot k \right) \cdot K_i - r_3 \cdot s_3 \cdot f \right).
$$
$$
\tag{5.22}
$$

This is an algebraic equation whose direct solution yields the unknown function $W(k)$.

We complete this section with a few remarks. Clearly, adding the diffusivity k as an extra-coordinate does not alter the overall computational complexity of the PGD procedure. This being said, the introduction of extra-coordinates may imply, for a given level of accuracy, the increase of the number of terms involved in the separated representation (5.3).

We have seen that, at each enrichment step, the construction of the new functional product (5.3) requires non-linear iterations. If Q_i denotes the number of iterations needed at enrichment step i, the total number of iterations involved in the construction of the PGD approximation is $Q = \sum_{i=1}^{N} Q_i$. In the above example, the entire procedure thus involves the solution of Q three-dimensional BVP's for the functions $X_i(\mathbf{x})$, Q one-dimensional IVP's for the functions $T_i(t)$, and Q algebraic diagonal systems for the functions $K_i(k)$. Thus, the complexity of the PGD procedure to compute the approximation (5.3) is essentially some tens of 3D steady-state BVP's, since the cost related to the 1D IVP's and algebraic problems are negligible in relative terms. With a standard numerical technique, however, one must solve for each particular value of the parameter k a 3D BVP at each time step. In usual applications, this often implies the computation of several millions of 3D solutions. Clearly, the CPU time savings afforded by the PGD can be of several orders of magnitude.

5.1.1 Numerical Example

Let us consider again a particular case of the transient diffusion equation

$$
\frac{\partial u}{\partial t} - k \cdot \frac{\partial^2 u}{\partial x^2} = f, \tag{5.23}
$$

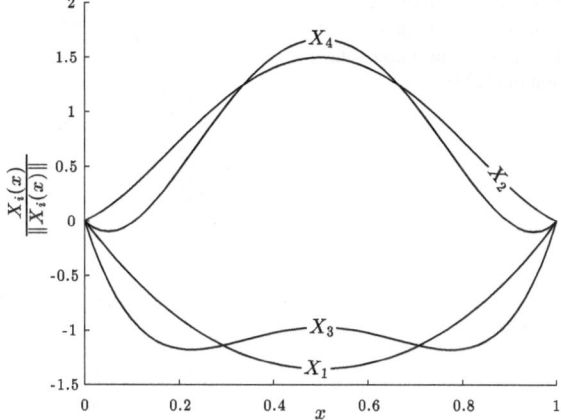

Fig. 5.1 Normalized functions $X_i(x)$ for $i = 1, \ldots, 4$ in the PGD solution of (5.23)

to be solved in the space-time-diffusivity domain $\Omega_x \times \Omega_t \times \Omega_k = (0, 1) \times (0, 0.1] \times [1, 5]$. For the the source term, we take $f = 1$. We specify homogeneous initial and boundary conditions, i.e. $u(x, t = 0) = u(x = 0, t) = u(x = 1, t) = 0$, which leads to the following analytical solution:

$$u_{\text{ex}}(x, t, k) = \frac{x(1 - x)}{2k} - \sum_{n \text{ odd}} \frac{4}{k \pi^3 n^3} \sin(n\pi x) \, e^{-k n^2 \pi^2 t} . \tag{5.24}$$

We seek a PGD solution of the form

$$u^N(x, t, k) = \sum_{i=1}^{N} X_i(x) \cdot T_i(t) \cdot K_i(k). \tag{5.25}$$

One immediately notices that the analytical solution cannot be expressed in this separated form. The unknown functions $X_i(x)$, $T_i(t)$ and $K_i(k)$ are sought on uniform grids with 61 points for the space domain, 151 points for the time domain and 101 points for the diffusivity domain. All spatial one-dimensional BVP's arising in the solution procedure are solved using second-order finite differences, while the IVP's are solved using the Backward Euler method. All integrals are evaluated numerically using the trapezoidal rule.

In Figs. 5.1, 5.2 and 5.3, we illustrate the normalized functions $X_i(x)$, $T_i(t)$ and $K_i(k)$ for $i = 1, \ldots, 4$. From Fig. 5.3, we can guess that considering $\frac{1}{k}$ as parametric coordinate instead of k might be a good idea.

To illustrate the convergence of the PGD solution to the analytical solution, we define the following quadratic error between the analytical solution $u_{\text{ex}}(x, t, k)$ and

Fig. 5.2 Normalized functions $T_i(t)$ for $i = 1, \ldots, 4$ in the PGD solution of (5.23)

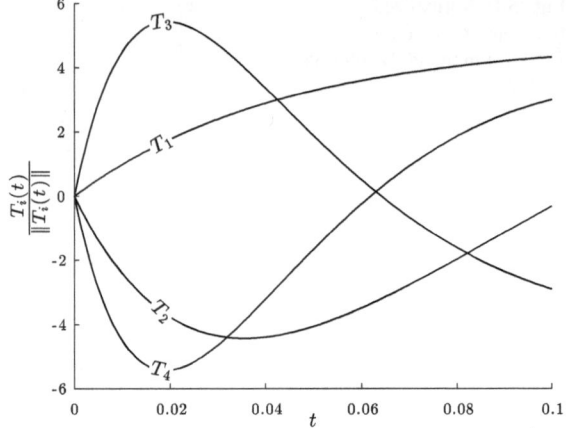

Fig. 5.3 Normalized functions $K_i(k)$ for $i = 1, \ldots, 4$ in the PGD solution of (5.23)

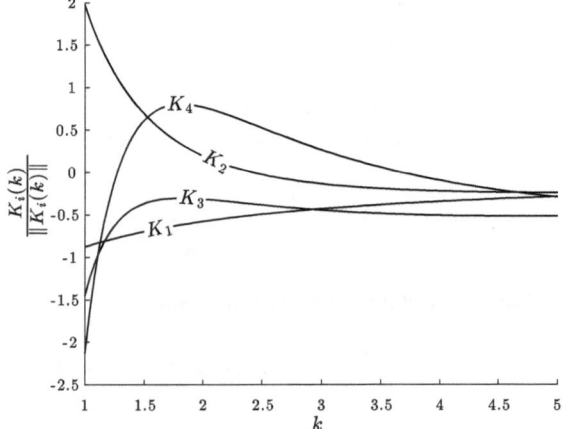

a numerical PGD solution $u^N(x, t, k)$ with N enrichment steps:

$$E(u^N) = \int_1^{\tilde{5}} \int_0^{\tilde{0.1}} \int_0^{\tilde{1}} \left(u_{\text{ex}}(x, t) - u^N(x, t) \right)^2 \, dx \cdot dt \cdot dk \qquad (5.26)$$

Here, the symbol $\tilde{\int}$ refers to a numerical integration carried out with the trapezoidal rule on the nodal values. In Fig. 5.4, we illustrate the global convergence of the PGD solution towards the analytical solution. The plain line is computed directly from the PGD solution $u^N(x, t, k)$, while the dashed line is computed after having applied the data post-compression described in Chap. 3. The advantage of the post-compression is obvious, as the compressed solution needs only five terms to reach the accuracy of the original twenty-five-term PGD solution.

Fig. 5.4 PGD error $E(u^N)$ as a function of the number N of enrichment steps. The plain line is computed directly from the PGD solution $u^N(x, t, k)$. The dashed line is computed after having applied the data post-compression described in Chap. 3

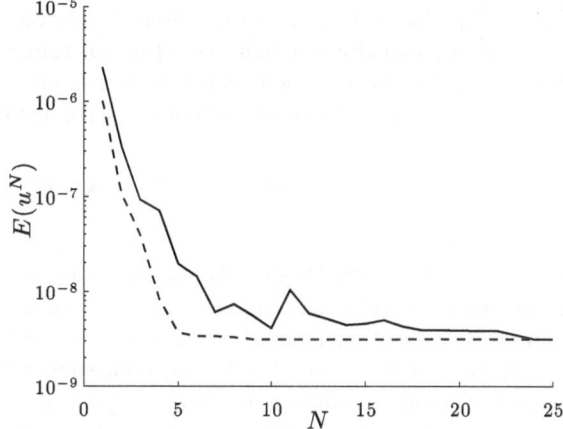

5.2 Boundary Conditions as Extra-Coordinates

In this section, we consider the steady-state, anisotropic diffusion equation in a three-dimensional domain Ω with Dirichlet conditions over part Γ_D of the boundary, and Neumann conditions over its complement Γ_N. We thus wish to solve

$$\nabla \cdot (\mathbf{K} \cdot \nabla u(\mathbf{x})) + f(\mathbf{x}) = 0, \tag{5.27}$$

for $\mathbf{x} \in \Omega \subset \mathcal{R}^3$, with the boundary conditions

$$\begin{cases} u(\mathbf{x} \in \Gamma_D) = u_g \\ (-\mathbf{K} \cdot \nabla u)|_{\mathbf{x} \in \Gamma_N} \cdot \mathbf{n} = \mathbf{q}_g \cdot \mathbf{n} = q_g \end{cases}. \tag{5.28}$$

Here, the symbol $\mathbf{K}$ denotes the diffusivity tensor, $\mathbf{n}$ is the unit outward normal vector at the boundary Γ_N, and the given data u_g and q_g are specified uniform values of the unknown field at Γ_D and the related flux at Γ_N, respectively.

5.2.1 Neumann Boundary Condition as Extra-Coordinate

First, imagine that we wish to know the model solution for $\mathcal{M}_q$ different values of the heat flux $q_g^1 < q_g^2 < \cdots < q_g^{\mathcal{M}_q}$ prescribed on the domain boundary Γ_N. In the classical approach, one would have to solve a new three-dimensional diffusion problem for *each* of the $\mathcal{M}_q$ values of the specified flux. If the space discretization involves $\mathcal{M}_\mathbf{x}$ nodes, we should thus compute and store $\mathcal{M}_\mathbf{x} \cdot \mathcal{M}_q$ discrete solution data.

The PGD framework offers a much more appealing alternative. Here, the prescribed flux q_g is considered as an extra-coordinate defined in the interval $\Omega_q =$

$[q_g^1, q_g^{\mathcal{M}_q}]$. One then solves a *single* four-dimensional diffusion equation to obtain the general parametric solution $u(\mathbf{x}, q)$ for *all* values of the prescribed flux in the interval Ω_q. Let us now briefly detail the procedure.

The parametric solution is sought in the separated form

$$u(\mathbf{x}, q_g) = \sum_{i=1}^{N} X_i(\mathbf{x}) \cdot Q_i(q_g). \tag{5.29}$$

In order to enforce the Dirichlet boundary condition $u(\mathbf{x} \in \Gamma_D) = u_g$, the simplest procedure consists in selecting the first functional couple $X_1(\mathbf{x}) \cdot Q_1(q_g)$ (first, or more if necessary) such that $u^1(\mathbf{x} \in \Gamma_D, q_g) = X_1(\mathbf{x} \in \Gamma_D) \cdot Q_1(q_g) = u_g$. Thus, the remaining terms $X_i(\mathbf{x})$, $i > 1$ of the finite sum (5.29) will be subjected to homogeneous Dirichlet conditions, i.e. $X_i(\mathbf{x} \in \Gamma_D) = 0$, $i > 1$. Alternatively, penalty or Lagrange multiplier formulations can be used to address Dirichlet conditions. Other possibilities are considered in [1].

In this PGD framework, the weighted residual form related to (5.27) and (5.28) reads

$$\int_{\Omega \times \Omega_q} \nabla u^* \cdot (\mathbf{K} \cdot \nabla u) \, d\mathbf{x} \cdot dq_g = - \int_{\Gamma_N \times \Omega_q} u^* \cdot q_g \, d\mathbf{x} \cdot dq_g + \int_{\Omega \times \Omega_q} u^* \cdot f(\mathbf{x}) \, d\mathbf{x} \cdot dq_g, \tag{5.30}$$

for all suitable test function u^*.

At enrichment step n, the solution is sought in the form

$$u^n(\mathbf{x}, q_g) = \sum_{i=1}^{n-1} X_i(\mathbf{x}) \cdot Q_i(q_g) + X_n(\mathbf{x}) \cdot Q_n(q_g) = u^{n-1}(\mathbf{x}, q_g) + X_n(\mathbf{x}) \cdot Q_n(q_g), \tag{5.31}$$

where $u^{n-1}(\mathbf{x}, q_g)$ is known. The new couple of unknown functions $X_n(\mathbf{x})$ and $Q_n(q_g)$ is computed by applying alternating direction iterations. In this particular example, each iteration involves two steps, with the test function u^* defined respectively as $X^*(\mathbf{x}) \cdot Q_n(q_g)$ and $X_n(\mathbf{x}) \cdot Q^*(q_g)$. We leave the details as an exercise.

Thus, the PGD solution (5.29) involves $N \cdot (\mathcal{M}_\mathbf{x} + \mathcal{M}_q)$ discrete data. If m is the total number of non-linear iterations, we need to solve a number m of three-dimensional problems for calculating the functions $X_i(\mathbf{x})$, $i = 1, \ldots, N$, as well as m one-dimensional problems for calculating the N functions $Q_i(q_g)$. The relative computing cost of these 1D problems is negligible. Thus, if $m < \mathcal{M}_q$, the PGD will proceed faster than the solution of the model for the different values of the parameter q_g^i, $i = 1, \ldots, \mathcal{M}_q$. From the point of view of data storage, the PGD is superior to the classical approach as soon as $N \cdot (\mathcal{M}_\mathbf{x} + \mathcal{M}_q) < \mathcal{M}_\mathbf{x} \cdot \mathcal{M}_q$.

When considering only one parameter as extra-coordinate the superiority of PGD with respect to standard procedures is not crucial, but as discussed previously and illustrated in what follows, when the number of extra-coordinates increases, the benefits can be impressive.

5.2.2 Dirichlet Boundary Condition as Extra-Coordinate

We now address the problem of computing the solution of the diffusion equation (5.27) for any value of the prescribed Dirichlet condition u_g in (5.28) in a certain interval $\Omega_u = [u_g^-, u_g^+]$.

For this purpose, we consider a continuous function $\varphi(\mathbf{x})$ such that $\varphi(\mathbf{x} \in \Gamma_D) = 1$. As proposed in [1], we introduce in (5.27) and (5.28) the following change of variables

$$u(\mathbf{x}) = v(\mathbf{x}) + u_g \cdot \varphi(\mathbf{x}), \tag{5.32}$$

which leads to

$$\nabla \cdot (\mathbf{K} \cdot \nabla v(\mathbf{x})) + u_g \cdot \nabla \cdot (\mathbf{K} \cdot \nabla \varphi(\mathbf{x})) + f(\mathbf{x}) = 0, \tag{5.33}$$

with the boundary conditions

$$\begin{cases} v(\mathbf{x} \in \Gamma_D) = 0 \\ (-\mathbf{K} \cdot \nabla v)|_{\mathbf{x} \in \Gamma_N} \cdot \mathbf{n} = u_g \cdot (\mathbf{K} \cdot \nabla \varphi)|_{\mathbf{x} \in \Gamma_N} \cdot \mathbf{n} + q_g \end{cases}. \tag{5.34}$$

We can now introduce u_g as an extra-coordinate. The parametric solution is then sought in the separated form

$$v(\mathbf{x}, u_g) = \sum_{i=1}^{N} X_i(\mathbf{x}) \cdot U_i(u_g). \tag{5.35}$$

In this PGD framework, the corresponding weighted residual form is given by

$$\int_{\Omega \times \Omega_u} \nabla v^* \cdot (\mathbf{K} \cdot \nabla v) \, d\mathbf{x} \cdot du_g$$

$$= - \int_{\Omega \times \Omega_u} \nabla v^* \cdot u_g \cdot (\mathbf{K} \cdot \nabla \varphi) \, d\mathbf{x} \cdot du_g + \int_{\Omega \times \Omega_u} v^* \cdot f(\mathbf{x}) \, d\mathbf{x} \cdot du_g$$

$$- \int_{\Gamma_N \times \Omega_u} v^* \cdot q_g \, d\mathbf{x} \cdot du_g - \int_{\Gamma_N \times \Omega_u} v^* \cdot u_g \cdot (\mathbf{K} \cdot \nabla \varphi) \cdot \mathbf{n} \, d\mathbf{x} \cdot du_g. \tag{5.36}$$

We leave it as an exercise to develop the details of the related alternating direction iterations.

5.2.3 Dirichlet and Neumann Boundary Conditions as Extra-Coordinates

In view of the above two sections, it readily follows that use of the weighted residual form

$$\int_{\Omega \times \Omega_u \times \Omega_q} \nabla v^* \cdot (\mathbf{K} \cdot \nabla v) \, d\mathbf{x} \cdot du_g \cdot dq_g$$

$$= -\int_{\Omega \times \Omega_u \times \Omega_q} \nabla v^* \cdot u_g \cdot (\mathbf{K} \cdot \nabla \varphi) \, d\mathbf{x} \cdot du_g \cdot dq_g + \int_{\Omega \times \Omega_u \times \Omega_q} v^* \cdot f(\mathbf{x}) \, d\mathbf{x} \cdot du_g \cdot dq_g$$

$$- \int_{\Gamma_N \times \Omega_u \times \Omega_q} v^* \cdot q_g \, d\mathbf{x} \cdot du_g \cdot dq_g - \int_{\Gamma_N \times \Omega_u \times \Omega_q} v^* \cdot u_g \cdot (\mathbf{K} \cdot \nabla \varphi) \cdot \mathbf{n} \, d\mathbf{x} \cdot du_g \cdot dq_g,$$

$$(5.37)$$

allows one to compute a parametric solution $u(\mathbf{x}, u_g, q_g)$ involving both Dirichlet and Neumann boundary conditions as extra-coordinates. The solution is then sought in the separated form

$$v(\mathbf{x}, u_g, q_g) = \sum_{i=1}^{N} X_i(\mathbf{x}) \cdot U_i(u_g) \cdot Q_i(q_g). \qquad (5.38)$$

5.2.4 Non-Constant Neumann Boundary Conditions

We now consider that in (5.28) the prescribed flux is no longer a constant, but rather a function of the space coordinate: $q_g = q_g(\mathbf{x})$, with $\mathbf{x} \in \Gamma_N$. We assume that $q_g(\mathbf{x})$ is defined on Γ_N for example by the finite element interpolation (other parametrizations are possible)

$$q_g(\mathbf{x}) = \sum_{k=1}^{S_q} Q_g^k \cdot \eta_k(\mathbf{x}), \qquad (5.39)$$

where Q_g^k are prescribed nodal fluxes at the nodal positions $\mathbf{x}_k \in \Gamma_N$, i.e. $Q_g^k = q_g(\mathbf{x}_k)$, and $\eta_k(\mathbf{x})$ are finite element shape functions.

Our goal is to compute the general parametric solution $u(\mathbf{x}, Q_g^1, \ldots, Q_g^{S_q})$ where the nodal fluxes Q_g^k can take *any* value in the intervals Ω_q^k, $k = 1, \ldots, S_q$.

An immediate extension of the weighted residual form (5.30) is thus given by

$$\int_{\Omega \times \Omega_q^1 \times \cdots \times \Omega_q^{S_q}} \nabla u^* \cdot (\mathbf{K} \cdot \nabla u) \, d\mathbf{x} \cdot dQ_g^1 \cdots dQ_g^{S_q}$$

$$= -\int_{\Gamma_N \times \Omega_q^1 \times \cdots \times \Omega_q^{S_q}} u^* \cdot \left(\sum_{k=1}^{S_q} Q_g^k \cdot \eta_k(\mathbf{x}) \right) d\mathbf{x} \cdot dQ_g^1 \cdots dQ_g^{S_q}$$

$$+ \int_{\Omega \times \Omega_q^1 \times \cdots \times \Omega_q^{S_q}} u^* \cdot f(\mathbf{x}) \, d\mathbf{x} \cdot dQ_g^1 \cdots dQ_g^{S_q}. \tag{5.40}$$

This allows one to compute the PGD solution in the separated form

$$u(\mathbf{x}, Q_g^1, \ldots, Q_g^{S_q}) = \sum_{i=1}^{N} X_i(\mathbf{x}) \cdot \prod_{j=1}^{S_q} G_i^j(Q_g^j). \tag{5.41}$$

For this class of problems, classical approaches fail in view of the curse of dimensionality as S_q increases, and use of the PGD strategy is an appealing alternative.

5.2.5 Non-Homogeneous Dirichlet Boundary Conditions

In this section, we consider the solution of the diffusion Eq. (5.27) with a non-homogeneous Dirichlet condition (5.28), i.e. $u_g = u_g(\mathbf{x})$ for $\mathbf{x} \in \Gamma_D$. We assume that $u_g(\mathbf{x})$ is defined on Γ_D by the finite element interpolation

$$u_g(\mathbf{x}) = \sum_{k=1}^{S_u} U_g^k \cdot \eta_k(\mathbf{x}), \tag{5.42}$$

where U_g^k are prescribed values at the nodal positions $\mathbf{x}_k \in \Gamma_D$, i.e. $U_g^k = u_g(\mathbf{x}_k)$, and $\eta_k(\mathbf{x})$ are finite element shape functions.

Our goal is to compute the general parametric solution $u(\mathbf{x}, U_g^1, \ldots, U_g^{S_u})$ where the nodal values U_g^k can take *any* value in the intervals Ω_u^k, $k = 1, \ldots, S_u$.

The PGD procedure is an extension of what has been detailed in Sect. 5.2.3. First, one defines the continuous functions $\varphi_k(\mathbf{x})$ such that $\varphi_k(\mathbf{x} \in \Gamma_D) = \eta_k(\mathbf{x})$. Then, the following change of variables

$$u(\mathbf{x}) = v(\mathbf{x}) + \sum_{k=1}^{S_u} U_g^k \cdot \varphi_k(\mathbf{x}), \tag{5.43}$$

leads to the extended version of the weighted residual form (5.36),

$$\int_{\Omega \times \Omega_u^1 \times \cdots \times \Omega_u^{S_u}} \nabla v^* \cdot (\mathbf{K} \cdot \nabla v) \, d\mathbf{x} \cdot dU_g^1 \cdots dU_g^{S_u}$$

$$= -\int_{\Omega \times \Omega_u^1 \times \cdots \times \Omega_u^{S_u}} \nabla v^* \cdot \left(\sum_{k=1}^{S_u} U_g^k \cdot (\mathbf{K} \cdot \nabla \varphi_k) \right) d\mathbf{x} \cdot dU_g^1 \cdots dU_g^{S_u}$$

$$+ \int_{\Omega \times \Omega_u^1 \times \cdots \times \Omega_u^{S_u}} v^* \cdot f(\mathbf{x}) \, d\mathbf{x} \cdot dU_g^1 \cdots dU_g^{S_u} - \int_{\Gamma_n \times \Omega_u^1 \times \cdots \times \Omega_u^{S_u}} v^* \cdot q_g \, d\mathbf{x} \cdot dU_g^1 \cdots dU_g^{S_u}$$

$$- \int_{\Gamma_n \times \Omega_u^1 \times \cdots \times \Omega_u^{S_u}} v^* \cdot \left(\sum_{k=1}^{S_u} U_g^k \cdot (\mathbf{K} \cdot \nabla \varphi_k) \cdot \mathbf{n} \right) d\mathbf{x} \cdot dU_g^1 \cdots dU_g^{S_u}. \qquad (5.44)$$

The solution is then sought in the separated form

$$v(\mathbf{x}, U_g^1, \cdots, U_g^{S_u}) = \sum_{i=1}^{N} X_i(\mathbf{x}) \cdot \prod_{j=1}^{S_u} F_i^j(U_g^j). \qquad (5.45)$$

5.3 Initial Conditions as Extra-Coordinates

We consider the transient diffusion equation for a homogeneous and isotropic medium,

$$\frac{\partial u}{\partial t} - k \cdot \Delta u = f, \qquad (5.46)$$

with uniform diffusivity k and source term f. We wish to integrate this equation for $t \in \Omega_t = (0, \Theta] \subset \mathcal{R}$ and $\mathbf{x} \in \Omega \subset \mathcal{R}^3$. The boundary and initial conditions are given by

$$\begin{cases} u(\mathbf{x} \in \Gamma_D) = u_g \\ (-k \cdot \nabla u)|_{\mathbf{x} \in \Gamma_N} \cdot \mathbf{n} = q_g \\ u(\mathbf{x}, t = 0) = u^0(\mathbf{x}) \end{cases} . \qquad (5.47)$$

We assume for example that the initial condition is defined in Ω by the finite element interpolation (there exist multiple possible parametrizations [2, 3])

$$u^0(\mathbf{x}) = \sum_{k=1}^{S_0} U_0^k \cdot \eta_k(\mathbf{x}), \qquad (5.48)$$

where U_0^k are prescribed values at nodes $\mathbf{x}_k \in \Omega$, i.e. $U_0^k = u^0(\mathbf{x}_k)$, and $\eta_k(\mathbf{x})$ are finite element shape functions.

Our goal is to compute the general parametric solution $u(\mathbf{x}, U_0^1, \ldots, U_0^{S_0})$ where the nodal values U_0^k can take *any* value in the intervals $\Omega_0^k = [(U_0^k)^-, (U_0^k)^+]$, for $k = 1, \ldots, S_0$.

We start with the associated weighted residual form,

$$\int_{\Omega} u^* \cdot \frac{\partial u}{\partial t} \, d\mathbf{x} + \int_{\Omega} \nabla u^* \cdot k \cdot \nabla u \, d\mathbf{x} = - \int_{\Gamma_N} u^* \cdot q_g \, d\mathbf{x} + \int_{\Omega} u^* \cdot f(\mathbf{x}) \, d\mathbf{x}, \quad (5.49)$$

wherein the Neumann condition appears explicitly as a natural boundary condition. In order to prescribe both the initial and the Dirichlet (essential) boundary conditions in the PGD framework, we first define the following functions:

$$\hat{u}^0(\mathbf{x}) = \begin{cases} u^0(\mathbf{x}) & \mathbf{x} \in \Omega \\ 0 & \mathbf{x} \in \Gamma \end{cases}, \qquad (5.50)$$

$$\Upsilon(t) = \begin{cases} 1 & t > 0 \\ 0 & t = 0 \end{cases}. \qquad (5.51)$$

Furthermore, we introduce a function $\varphi(\mathbf{x})$ continuous in Ω that satisfies the Dirichlet conditions,

$$\varphi(\mathbf{x} \in \Gamma_D) = u_g. \qquad (5.52)$$

We could then define the function $\Sigma(\mathbf{x}, t)$ expressed in the separated form

$$\Sigma(\mathbf{x}, t) = \hat{u}^0(\mathbf{x}) + \varphi(\mathbf{x}) \cdot \Upsilon(t). \qquad (5.53)$$

This new function satisfies both the initial and Dirichlet conditions. The functions $\hat{u}^0$ and $\Upsilon(t)$, however, are not sufficiently regular to be used in the weighted residual formulation of the problem. A direct regularization consists in defining these functions at the nodal positions and then define interpolations with the required regularity. Thus, the discrete counterpart of functions $\hat{u}^0$ and $\Upsilon(t)$ are given by

$$\hat{u}^0(\mathbf{x}_k) = \begin{cases} u^0(\mathbf{x}_k) & \mathbf{x}_k \in \Omega \\ 0 & \mathbf{x}_k \in \Gamma \end{cases}, \qquad (5.54)$$

and

$$\Upsilon(t_l) = \begin{cases} 1 & t_l > 0 \\ 0 & t_l = 0 \end{cases}, \qquad (5.55)$$

with $k = 1, \ldots, 1, \ldots, \mathcal{M}_\mathbf{x}$ and $l = 1, \ldots, \mathcal{M}_t$. Standard finite element interpolation is then used to define the functions $\hat{u}^0(\mathbf{x})$ and $\Upsilon(t)$ everywhere from their nodal values given by (5.54) and (5.55).

We can now apply the change of variable

$$u(\mathbf{x}, t) = v(\mathbf{x}, t) + \Sigma(\mathbf{x}, t) = v(\mathbf{x}, t) + \hat{u}^0(\mathbf{x}) + \varphi(\mathbf{x}) \cdot \Upsilon(t), \qquad (5.56)$$

and the weighted residual form (5.49) becomes

$$\int_{\Omega} v^* \cdot \frac{\partial v}{\partial t} \, d\mathbf{x} + \int_{\Omega} \nabla v^* \cdot k \cdot \nabla v \, d\mathbf{x} = - \int_{\Omega} v^* \cdot \varphi \cdot \frac{\partial \Upsilon}{\partial t} \, d\mathbf{x}$$

$$- \int_{\Omega} \nabla v^* \cdot k \cdot \nabla \hat{u}^0 \, d\mathbf{x} - \int_{\Gamma_N} v^* \cdot k \cdot \nabla \hat{u}^0 \cdot \mathbf{n} \, dx$$

$$- \int_{\Gamma_N} v^* \cdot q_g \, d\mathbf{x} - \int_{\Omega} \nabla v^* \cdot k \cdot \Upsilon \cdot \nabla \varphi \, dx$$

$$- \int_{\Gamma_N} v^* \cdot k \cdot \Upsilon \cdot \nabla \varphi \cdot \mathbf{n} \, dx + \int_{\Omega} v^* \cdot f(\mathbf{x}) \, dx. \tag{5.57}$$

The PGD parametric solution $u(\mathbf{x}, U_0^1, \ldots, U_0^{S_0})$ is sought in the form

$$u(\mathbf{x}, U_0^1, \ldots, U_0^{S_0}) = \sum_{i=1}^{N} X_i(\mathbf{x}) \cdot \prod_{j=1}^{S_0} \mathcal{U}_i^j (U_0^j). \tag{5.58}$$

In view of (5.48) and (5.57), it is obtained from the extended weighted residual form

$$\int_{\Omega \times \Omega_0^1 \times \cdots \times \Omega_0^{S_0}} v^* \cdot \frac{\partial v}{\partial t} \, d\mathbf{x} \cdot dU_0^1 \cdots dU_0^{S_0} + \int_{\Omega \times \Omega_0^1 \times \cdots \times \Omega_0^{S_0}} \nabla v^* \cdot k \cdot \nabla v \, d\mathbf{x} \cdot dU_0^1 \cdots dU_0^{S_0}$$

$$= - \int_{\Omega \times \Omega_0^1 \times \cdots \times \Omega_0^{S_0}} v^* \cdot \varphi \cdot \frac{\partial \Upsilon}{\partial t} \, d\mathbf{x} \cdot dU_0^1 \cdots dU_0^{S_0}$$

$$- \int_{\Omega \times \Omega_0^1 \times \cdots \times \Omega_0^{S_0}} \nabla v^* \cdot k \cdot \left(\sum_{k=1}^{S_0} U_0^k \cdot \nabla \eta_k(\mathbf{x}) \right) d\mathbf{x} \cdot dU_0^1 \cdots dU_0^{S_0}$$

$$- \int_{\Gamma_N \times \Omega_0^1 \times \cdots \times \Omega_0^{S_0}} v^* \cdot k \cdot \left(\sum_{k=1}^{S_0} U_0^k \cdot \eta_k(\mathbf{x}) \cdot \mathbf{n} \right) d\mathbf{x} \cdot dU_0^1 \cdots dU_0^{S_0}$$

$$- \int_{\Gamma_N \times \Omega_0^1 \times \cdots \times \Omega_0^{S_0}} v^* \cdot q_g \, d\mathbf{x} \cdot dU_0^1 \cdots dU_0^{S_0}$$

$$- \int_{\Omega \times \Omega_0^1 \times \cdots \times \Omega_0^{S_0}} \nabla v^* \cdot k \cdot \Upsilon \cdot \nabla \varphi \, d\mathbf{x} \cdot dU_0^1 \cdots dU_0^{S_0}$$

$$- \int_{\Gamma_N \times \Omega_0^1 \times \cdots \times \Omega_0^{S_0}} v^* \cdot k \cdot \Upsilon \cdot \nabla \varphi \cdot \mathbf{n} \, d\mathbf{x} \cdot dU_0^1 \cdots dU_0^{S_0}$$

$$+ \int_{\Omega \times \Omega_0^1 \times \cdots \times \Omega_0^{S_0}} v^* \cdot f(\mathbf{x}) \, d\mathbf{x} \cdot dU_0^1 \cdots dU_0^{S_0}. \tag{5.59}$$

5.4 Computational Domain as Extra-Coordinates

The parameters defining the computational domain can also be considered as extra-coordinates in the PGD framework. This allows us to compute a general parametric solution for an *ensemble* of computational domains.

Consider for example the transient one-dimensional heat equation

$$\frac{\partial u}{\partial t} - k \cdot \frac{\partial^2 u}{\partial x^2} = f, \tag{5.60}$$

defined in the domain $\Omega = (0, L)$ over a time interval $I_t = (0, \Theta]$. The source term f and diffusivity k are uniform, and we specify homogeneous initial and boundary conditions, namely $u(x = 0, t) = u(x = L, t) = u(x, t = 0) = 0$.

We wish to compute by means of the PGD the general parametric solution $u(x, t; L, \Theta)$ for *all* domain length L in the given interval $[L^-, L^+]$ and for *all* time duration Θ in the given interval $[\Theta^-, \Theta^+]$.

The starting point is the associated weighted residual formulation of the problem,

$$\int_{\Omega \times \Omega_t} u^* \cdot \frac{\partial u}{\partial t} \, dx \cdot dt = -k \cdot \int_{\Omega \times \Omega_t} \frac{\partial u^*}{\partial x} \cdot \frac{\partial u}{\partial x} \, dx \cdot dt + \int_{\Omega \times \Omega_t} u^* \cdot f \, dx \cdot dt. \tag{5.61}$$

We must now introduce the extra-coordinates L and Θ explicitly in (5.61). To do so, consider the coordinate transformation

$$\begin{cases} t = \tau \cdot \Theta \ \tau \in [0, 1] \\ x = \lambda \cdot L \ \lambda \in [0, 1] \end{cases}. \tag{5.62}$$

The weighted residual form (5.61) then becomes

$$\int_{[0,1]^2} u^* \cdot \frac{\partial u}{\partial \tau} \cdot L \, d\lambda \cdot d\tau = -k \cdot \int_{[0,1]^2} \frac{\partial u^*}{\partial \lambda} \cdot \frac{\partial u}{\partial \lambda} \cdot \frac{\Theta}{L} \, d\lambda \cdot d\tau + \int_{[0,1]^2} u^* \cdot f \cdot L \cdot \Theta \, d\lambda \cdot d\tau. \tag{5.63}$$

The PGD procedure can then be implemented in the usual manner, yielding the parametric solution expressed in terms of the transformed coordinates,

$$u(\lambda, \tau, L, \Theta) = \sum_{i=1}^{N} X_i(\lambda) \cdot T_i(\tau) \cdot \mathcal{L}_i(L) \cdot \mathcal{T}_i(\Theta). \tag{5.64}$$

References

1. D. Gonzalez, A. Ammar, F. Chinesta, E. Cueto, Recent advances in the use of separated representations. Int. J. Numer. Meth. Eng. **81**(5), 637–659 (2010)

2. A. Ammar, F. Chinesta, E. Cueto, M. Doblare, Proper generalized decomposition of time-multiscale models. Int. J. Numer. Meth. Eng. **90**(5), 569–596 (2012)
3. D. Gonzalez, F. Masson, F. Poulhaon, A. Leygue, E. Cueto, F. Chinesta, Proper generalized decomposition based dynamic data-driven inverse identification. Math. Comput. Simul. **82**(9), 1677–1695 (2012)

Chapter 6
Advanced Topics

Abstract This chapter addresses some advanced topics whose development remains work in progress. The first topic concerns the efficient treatment of non-linear models where standard strategies can fail for high-dimensional problems. The second topic concerns the use of advective stabilization when the involved fields are approximated in a separated form. Finally, we introduce a discrete form of the PGD solver, the one that we consider in computer implementations, that is then extended for considering a separated representation constructor based on residual minimization. Residual minimization is particularly suitable for addressing non-symmetric differential operators, for which the standard procedure described in the previous chapters can be inefficient (slow convergence and non-optimal representations).

Keywords Advection-Diffusion equation · Discrete representation · Nonlinear models · Residual minimization

In this final chapter, we address three advanced topics of major interest that require a specific approach in the context of the PGD: solution of non-linear models, stabilization schemes for convection-dominated problems, and treatment of non-symmetric differential operators.

6.1 Non-Linear Models

Non-linear models abound in many applications and their solution requires appropriate linearization schemes. In [1, 2], we have extended standard linearization procedures to the PGD framework. We start here by considering such procedures in order to point out their inherent limitations. A powerful alternative approach that we are developing at present is discussed next.

F. Chinesta et al., *The Proper Generalized Decomposition for Advanced Numerical Simulations*, SpringerBriefs in Applied Sciences and Technology, DOI: 10.1007/978-3-319-02865-1_6, © The Author(s) 2014

6.1.1 PGD Strategies Based on Standard Linearization Schemes

Following [1, 2], we consider the non-linear transient diffusion problem

$$\frac{\partial u}{\partial t} - k \cdot \Delta u = -u^2 + f(\mathbf{x}, t), \tag{6.1}$$

with homogeneous boundary and initial conditions

$$\begin{cases} u(\mathbf{x} \in \Gamma, t \in (0, T_{max}]) = 0 \\ u(\mathbf{x} \in \Omega, t = 0) = 0 \end{cases}, \tag{6.2}$$

and $\Omega \subset \mathbb{R}^D$, $D \geq 1$, $T_{max} > 0$.

We wish to obtain an approximated solution of this problem by means of the PGD. At this stage, two standard linearization schemes can be envisaged to deal with the non-linear term u^2 in (6.1), namely a fixed-point Picard scheme and Newton's method.

With the Picard scheme, we seek the solution in the usual separated form

$$u(\mathbf{x}, t) = \sum_{i=1}^{N} X_i(\mathbf{x}) \cdot T_i(t). \tag{6.3}$$

At a particular enrichment step n, the $n - 1$ first terms of this sum are known and the next functional product $X_n(\mathbf{x}) \cdot T_n(t)$ must be computed to yield the updated approximation:

$$u^n(\mathbf{x}, t) = u^{n-1}(\mathbf{x}, t) + X_n(\mathbf{x}) \cdot T_n(t) = \sum_{i=1}^{n-1} X_i(\mathbf{x}) \cdot T_i(t) + X_n(\mathbf{x}) \cdot T_n(t). \tag{6.4}$$

In the Picard scheme, the non-linear term u^2 in (6.1) is evaluated by means of the solution at the previous enrichment step u^{n-1},

$$u^2 \approx \left(\sum_{i=1}^{n-1} X_i(\mathbf{x}) \cdot T_i(t) \right)^2 \tag{6.5}$$

As described previously, the alternating direction strategy then proceeds by calculating $X_n^p(\mathbf{x})$ from the just-computed $T_n^{p-1}(t)$, and then updating $T_n^p(\mathbf{x})$ from $X_n^p(\mathbf{x})$, until convergence is reached.

With Newton's method, we first linearize the governing equation and then apply the PGD at each Newton iteration.

Thus, at Newton iteration j, the updated solution is given by

$$u^{j+1} = u^j + \widetilde{u}, \tag{6.6}$$

where the increment $\widetilde{u}$ is the solution of the linearized problem

$$\frac{\partial \widetilde{u}}{\partial t} - k \cdot \Delta \widetilde{u} + 2 u^j \, \widetilde{u} = -\mathcal{R}(u^j), \tag{6.7}$$

with the residual $\mathcal{R}(u^j)$ given by

$$\mathcal{R}(u^j) = \left(\frac{\partial u^j}{\partial t} - k \cdot \Delta u^j + (u^j)^2 - f(\mathbf{x}, t) \right). \tag{6.8}$$

This linear problem is then solved by means of the PGD algorithm. Thus, $\widetilde{u}$ is obtained in the separated form

$$\widetilde{u}(\mathbf{x}, t) = \sum_{i=1}^{\widetilde{N}} \widetilde{X}_i(\mathbf{x}) \cdot \widetilde{T}_i(t), \tag{6.9}$$

and the usual enrichment procedure is applied.

6.1.2 Discussion

In the context of the PGD, both Picard and Newton procedures converge but no significant difference in the number of required iterations is noticed. The Newton strategy converges slightly faster, but the coupling of the outer Newton iteration with the enrichment procedure prevents the quadratic convergence rate expected from a Newton scheme. As discussed in [3], even when the exact solution can be represented by a single functional product, i.e.

$$u^{ex}(\mathbf{x}, t) = X^{ex}(\mathbf{x}) \cdot T^{ex}(t), \tag{6.10}$$

the Newton iterations produce, by construction, a solution in separated form that involves a number of functional products proportional to the number of Newton iterations.

The Picard scheme can be improved rather easily by evaluating the non-linear term u^2 in (6.1) at iteration p of enrichment step n,

$$u^2 \approx \left(u^{n-1} + X_n^{p-1}(\mathbf{x}) \cdot T_n^{p-1}(t) \right)^2. \tag{6.11}$$

When the exact solution can be represented by a single functional product, the improved Picard scheme converges after computing the first functional couple [2].

In that sense, the improved Picard scheme is optimal but the overall computing time is similar to that of the standard fixed point or the Newton strategy.

The main difficulty related to the use of standard linearization schemes lies in the evaluation of the non-linear terms (u^2 in our example), which can become inordinately expensive when the number of terms in the separated representation, the problem's dimension D increases and/or the power of the non-linear term increases (note that general non-linear functions can be approximated by using a Taylor polynomial expansion).

A technique based on the Asymptotic Numerical Method—ANM—[4, 5] was proposed in [3]. Despite some improvements the just referred issues persisted. LATIN based solvers intensively developed by P. Ladevèze and his collaborators are not at present able to address multi-parametric non-linear models.

An appealing alternative strategy is based on the interpolation of the non-linear term as proposed in [6, 7]. This strategy was extended to the PGD method in [8] and it is described next.

6.1.3 Interpolation of the Non-Linear Term

We consider an extension of the previous problem,

$$\frac{\partial u}{\partial t} - k \cdot \Delta u = \mathcal{L}(u) + f(\mathbf{x}, t), \tag{6.12}$$

where $\mathcal{L}(u)$ is now a general non-linear function of u. Homogeneous initial and boundary conditions are again specified.

We first seek the solution $u^0(\mathbf{x}, t)$ of the associated linear problem

$$\frac{\partial u^0}{\partial t} - k \cdot \Delta u^0 = f(\mathbf{x}, t), \tag{6.13}$$

in the PGD separated form

$$u^0(\mathbf{x}, t) = \sum_{i=1}^{N^0} X_i^0(\mathbf{x}) \cdot T_i^0(t). \tag{6.14}$$

We thus obtain the reduced approximation basis $\mathcal{B}^0 = \{\tilde{X}_1^0 \cdot \tilde{T}_1^0, \ldots, \tilde{X}_{N^0}^0 \cdot \tilde{T}_{N^0}^0\}$ that contains the normalized functions $\tilde{X}_i^0 = \frac{X_i^0}{\|X_i^0\|}$ and $\tilde{T}_i^0 = \frac{T_i^0}{\|T_i^0\|}$.

An interpolation of the non-linear function $\mathcal{L}(u)$ can now be defined by means of the basis $\mathcal{B}^0$. For this purpose, we consider N^0 points $(\mathbf{x}_j^0, t_j^0)$, $j = 1, \ldots, N^0$, and we enforce that

$$\mathcal{L}(u^0(\mathbf{x}_j^0, t_j^0)) = \sum_{i=1}^{N^0} \xi_i^0 \cdot \tilde{X}_i^0(\mathbf{x}_j^0) \cdot \tilde{T}_i^0(t_j^0), \quad j = 1, \ldots, N^0. \tag{6.15}$$

This yields a linear algebraic system for the N^0 coefficients ξ_i^0. Once these coefficients ξ_i^0 have been computed, the interpolation $\mathcal{L}^0$ of the non-linear term is fully defined:

$$\mathcal{L}^0 \equiv \mathcal{L}(u^0(\mathbf{x}, t)) \approx \sum_{i=1}^{N^0} \xi_i^0 \cdot \tilde{X}_i^0(\mathbf{x}) \cdot \tilde{T}_i^0(t). \tag{6.16}$$

With $\mathcal{L}$ replaced by $\mathcal{L}^0$, the original non-linear problem then yields a linear problem involving $u^1(\mathbf{x}, t)$,

$$\frac{\partial u^1}{\partial t} - k \cdot \Delta u^1 = \mathcal{L}^0 + f(\mathbf{x}, t). \tag{6.17}$$

There are various ways of computing the solution $u^1(\mathbf{x}, t)$ [8].

Let us define N^1 as the number of approximation functions involved in the separated representation of $u^1(\mathbf{x}, t)$ and the associated reduced approximation basis $\mathcal{B}^1 = \{\tilde{X}_1^1 \cdot \tilde{T}_1^1, \ldots, \tilde{X}_{N^1}^1 \cdot \tilde{T}_{N^1}^1\}$. The non-linear term is then again interpolated by means of N^1 points $(\mathbf{x}_j^1, t_j^1)$, $j = 1, \ldots, N^1$ as done previously,

$$\mathcal{L}^1 \equiv \mathcal{L}(u^1(\mathbf{x}, t)) \approx \sum_{i=1}^{N^1} \xi_i^1 \cdot \tilde{X}_i^1(\mathbf{x}) \cdot \tilde{T}_i^1(t). \tag{6.18}$$

With $\mathcal{L}$ replaced by $\mathcal{L}^1$, the original non-linear problem thus leads to a linear problem for the function $u^2(\mathbf{x}, t)$. This procedure is repeated until reaching convergence.

The only issue that deserves additional comments is the choice of interpolation points $(\mathbf{x}_j^k, t_j^k)$, $j = 1, \ldots, N^k$, at iteration k of the non-linear solver.

At iteration k, the reduced approximation basis is given by

$$\mathcal{B}^k = \{\tilde{X}_1^k \cdot \tilde{T}_1^k, \ldots, \tilde{X}_{N^k}^k \cdot \tilde{T}_{N^k}^k\}. \tag{6.19}$$

Following [6, 7], we consider

$$(\mathbf{x}_1^k, t_1^k) = argmax_{\mathbf{x}, t} |\tilde{X}_1^k(\mathbf{x}) \cdot \tilde{T}_1^k(t)|. \tag{6.20}$$

We then compute d_1 from

$$d_1 \cdot \tilde{X}_1^k(\mathbf{x}_1^k) \cdot \tilde{T}_1^k(t_1^k) = \tilde{X}_2^k(\mathbf{x}_1^k) \cdot \tilde{T}_2^k(t_1^k). \tag{6.21}$$

Defining $r_2^k(\mathbf{x}, t)$ as

$$r_2^k(\mathbf{x}, t) = \tilde{X}_2^k(\mathbf{x}) \cdot \tilde{T}_2^k(t) - d_1 \cdot \tilde{X}_1^k(\mathbf{x}) \cdot \tilde{T}_1^k(t), \tag{6.22}$$

we compute the point $(\mathbf{x}_2^k, t_2^k)$ according to

$$(\mathbf{x}_2^k, t_2^k) = argmax_{\mathbf{x},t} |r_2^k(\mathbf{x}, t)|. \tag{6.23}$$

By construction, $r_2^k(\mathbf{x}_1^k, t_1^k) = 0$ and we can thus ensure $(\mathbf{x}_2^k, t_2^k) \neq (\mathbf{x}_1^k, t_1^k)$.

The procedure is generalized for obtaining the other interpolation points $(\mathbf{x}_j^k, t_j^k)$, $j \leq k$. Thus, we consider

$$r_j^k(\mathbf{x}, t) = \tilde{X}_j^k(\mathbf{x}) \cdot \tilde{T}_j^k(t) - \sum_{i=1}^{j-1} d_i \cdot \tilde{X}_i^k(\mathbf{x}) \cdot \tilde{T}_i^k(t), \tag{6.24}$$

and compute the interpolation point $(\mathbf{x}_j^k, t_j^k)$ as

$$(\mathbf{x}_j^k, t_j^k) = argmax_{\mathbf{x},t} |r_j^k(\mathbf{x}, t)|. \tag{6.25}$$

The coefficients $d_1, \ldots, d_{j-1}$ must be such that $(\mathbf{x}_j^k, t_j^k) \neq (\mathbf{x}_i^k, t_i^k)$, $\forall i < j \leq k$. For this purpose, we enforce that the residual $r_j^k(\mathbf{x}, t)$ vanishes at each location $(\mathbf{x}_i^k, t_i^k)$ with $i < j$, that is:

$$r_j^k(\mathbf{x}_l^k, t_l^k) = 0 = \tilde{X}_j^k(\mathbf{x}_l^k) \cdot \tilde{T}_j^k(t_l^k) - \sum_{i=1}^{j-1} d_i \cdot \tilde{X}_i^k(\mathbf{x}_l^k) \cdot \tilde{T}_i^k(t_l^k), \; l = 1, \ldots, j-1 \tag{6.26}$$

This is a linear algebraic system whose solution yields the coefficients $d_1, \ldots, d_{j-1}$.

6.1.4 Numerical Example

We illustrate here the differences between the Picard and Newton linearization strategies for the PGD applied to the one-dimensional problem (6.1) with $k = 0.1$, $T_{max} = 1$, $f = 1$ and $\Omega = (0, 1)$. A reference solution u_{ref} computed with a finite difference scheme and the standard Matlab® ODE solver is compared to the solutions obtained using both the Picard and the Newton linearization for different numbers of iterations and different numbers of enrichment steps per iteration. We define $u_{k,l}^P$ (resp. $u_{k,l}^N$) as the solution obtained after k Picard (resp. Newton) iterations where l functional products are computed with the PGD at each iteration. The error level of each solution is evaluated as the following quadratic error with respect to the reference solution, integrated over space and time using the trapezoidal rule:

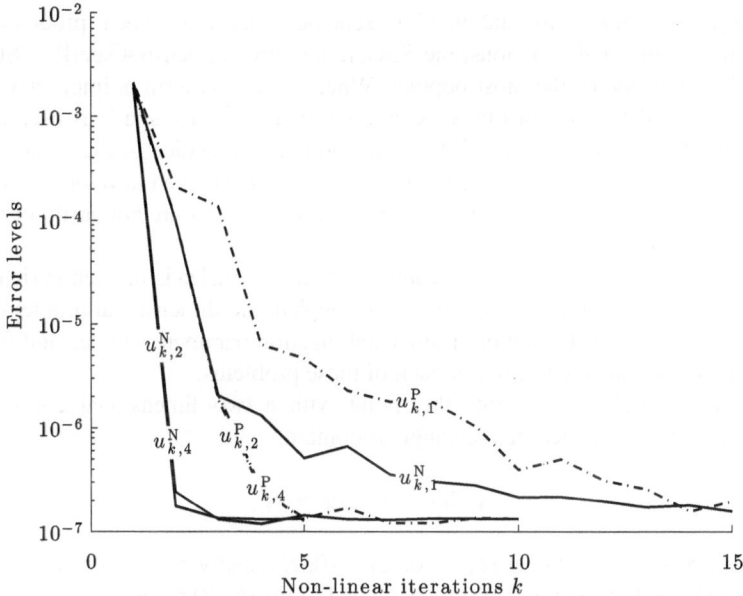

Fig. 6.1 Convergence of the PGD solutions towards the reference solution as a function of the number of Picard (*dashed lines*) and Newton (*plain lines*) iterations

$$E^{(N,P)} = \int_0^{\tilde{1}} \int_0^{\tilde{1}} \left(u_{\text{ref}} - u_{k,l}^{(N,P)} \right)^2 \, dx \cdot dt \,, \tag{6.27}$$

where (N, P) represents N or P.

In Fig. 6.1 we observe that for both linearization strategies, the computation of only one functional product ($l = 1$) per iteration is sub-optimal. For higher values of l, both methods converge well but they do not benefit much from increasing l. The Newton approach is always superior, especially for the first iterations. One should however keep in mind that Newton iterations are more expensive as they require the update of the tangent operator. With the naive Matlab® implementation used in this numerical example, the Newton linearization is about twice slower than the Picard scheme.

6.2 Convective Stabilization

It is well-known that standard finite element (Galerkin) methods are not suitable for convection-dominated problems, since they yield unstable, spuriously oscillating solutions [9].

Among the numerous stabilization schemes that have been proposed for convection-dominated equations, the Streamline-Upwind/Petrov-Galerkin (SUPG) method [10] is one of the most popular. When a reaction term is important, Sub-Grid Scale (SGS) techniques have been advocated to reduce spurious oscillations [11]. An inherent difficulty of all these methods is the choice of the stabilization parameter. In fact, algebraic or asymptotic analyses have been developed for one-dimensional problems, but optimal stabilization parameters are not easily obtained in higher dimensions.

The PGD is well suited in this context. Indeed, as it leads at each enrichment step to the solution of a large number of decoupled one-dimensional problems, the PGD allows for the selection of optimal stabilization parameters to account for the convection-dominated character of each of these problems.

Following [12], we illustrate this point with a two-dimensional convection-diffusion equation defined in a rectangular domain:

$$\mathbf{v} \cdot \nabla u - k \cdot \Delta u = f, \tag{6.28}$$

with $u(\mathbf{x})$, $\mathbf{x} \in \Omega$, $\Omega = \Omega_x \times \Omega_y = (0, L) \times (0, H)$, and $\mathbf{v}$ constant in Ω.

The weighted residual integral form related to Eq. (6.28) reads

$$\int_{\Omega_x \times \Omega_y} u^* \cdot \left(v_x \cdot \frac{\partial u}{\partial x} + v_y \cdot \frac{\partial u}{\partial y} - k \cdot \frac{\partial^2 u}{\partial x^2} - k \cdot \frac{\partial^2 u}{\partial y^2} - f \right) dx \cdot dy = 0, \tag{6.29}$$

for all suitable test functions u^*.

The PGD solution is sought in the separated form

$$u(x, y) = \sum_{i=1}^{N} X_i(x) \cdot Y_i(y). \tag{6.30}$$

At enrichment step n of the PGD algorithm, we have already computed the approximation

$$u^{n-1}(x, y) = \sum_{i=1}^{n-1} X_i(x) \cdot Y_i(y). \tag{6.31}$$

and we wish to obtain the next one, i.e.

$$u^n(x, y) = u^{n-1}(x, y) + X_n(x) \cdot Y_n(y) = \sum_{i=1}^{n-1} X_i(x) \cdot Y_i(y) + X_n(x) \cdot Y_n(y). \tag{6.32}$$

An alternating direction iterative scheme is then used to solve the non-linear problem for $X_n(x)$ and $Y_n(y)$. At iteration p, we must compute $X_n^p(x)$ from $Y_n^{p-1}(y)$, and then $Y_n^p(x)$ from $X_n^p(x)$. Let us detail the first step. At this stage, the approximation reads

$$u^n(x, y) = \sum_{i=1}^{n-1} X_i(x) \cdot Y_i(y) + X_n^p(x) \cdot Y_n^{p-1}(y), \tag{6.33}$$

where all functions except $X_n^p(x)$ are known. Selecting $u^*(x, y) = X_n^*(x) \cdot Y_n^{p-1}(y)$ for the test function and introducing (6.33) into (6.29), we obtain

$$\int_{\Omega_x \times \Omega_y} X_n^* \cdot Y_n^{p-1} \cdot$$

$$\left(v_x \cdot \frac{dX_n^p}{dx} \cdot Y_n^{p-1} + v_y \cdot X_n^p \cdot \frac{dY_n^{p-1}}{dy} - k \cdot \frac{d^2 X_n^p}{dx^2} \cdot Y_n^{p-1} - k \cdot X_n^p \cdot \frac{d^2 Y_n^{p-1}}{dy^2} \right) dx \cdot dy$$

$$= -\int_{\Omega_x \times \Omega_y} X_n^* \cdot Y_n^{p-1} \cdot$$

$$\sum_{i=1}^{n-1} \left(v_x \cdot \frac{dX_i}{dx} \cdot Y_i + v_y \cdot X_i \cdot \frac{dY_i}{dy} - k \cdot \frac{d^2 X_i}{dx^2} \cdot Y_i - k \cdot X_i \cdot \frac{d^2 Y_i}{dy^2} \right) dx \cdot dy$$

$$+ \int_{\Omega_x \times \Omega_y} X_n^* \cdot Y_n^{p-1} \cdot f \, dx \cdot dy. \tag{6.34}$$

Integration over Ω_y then yields the weighted residual form of a one-dimensional problem for the unknown function $X_n^p(x)$, which involves known coefficients α^x, β^x, γ^x, δ_i^x, χ_i^x, υ_i^x and ξ^x whose definition is left as an exercise:

$$\int_{\Omega_x} X_n^* \cdot \left(\alpha^x \cdot \frac{d^2 X_n^p}{dx^2} + \beta^x \cdot \frac{dX_n^p}{dx} + \gamma^x \cdot X_n^p \right) dx$$

$$- -\int_{\Omega_x} X_n^* \cdot \sum_{i=1}^{n-1} \left(\delta_i^x \cdot \frac{d^2 X_i}{dx^2} + \chi_i^x \cdot \frac{dX_i}{dx} + \upsilon_i^x \cdot X_l \right) dx + \int_{\Omega_x} X_n^* \cdot \xi^x \, dx. \tag{6.35}$$

The corresponding strong form is a one-dimensional convection-diffusion-reaction equation with a source term, for which quasi-optimal stabilization methods exist [9]. The interested reader can refer to [12] for a deeper analysis of this topic and numerical tests proving the performance of this approach, in particular for high-dimensional problems.

6.2.1 Numerical Example

To illustrate the above considerations, we consider the problem (6.28) with $L = H = 2$, $v_x = v_y = 1$, $k = 10^{-3}$ and the following expression for the source term $f(x, y)$:

$$f(x, y) = 10\, e^{-100(x-1)^2}\, e^{-100(y-1)^2}\,. \tag{6.36}$$

The following Dirichlet boundary conditions are applied:

$$\begin{cases} u(0, y) = u(x, 0) = 0 \\ u(x, 2) = \frac{x}{2} \\ u(2, y) = \frac{y}{2} \end{cases}, \tag{6.37}$$

which generate a boundary layer at the outflow boundary. This problem is a challenge for the PGD as the solution changes mostly along the diagonal ($x = y$) of the domain and is therefore poorly suited for a (x, y) separated representation. The PGD solution is computed using a finite difference method along each dimension. For all one-dimensional problems arising in the solution procedure, the convective term is stabilized using upwind finite differences. For more accurate stabilizations the interested reader can refer to [12]. In Fig. 6.2, we compare the solution with and without stabilization. Without stabilization, one observes the typical oscillations generated by the lack on convective stabilization, while the stabilized solution is smooth and correctly captures the boundary layer near the top and right boundaries of the domain.

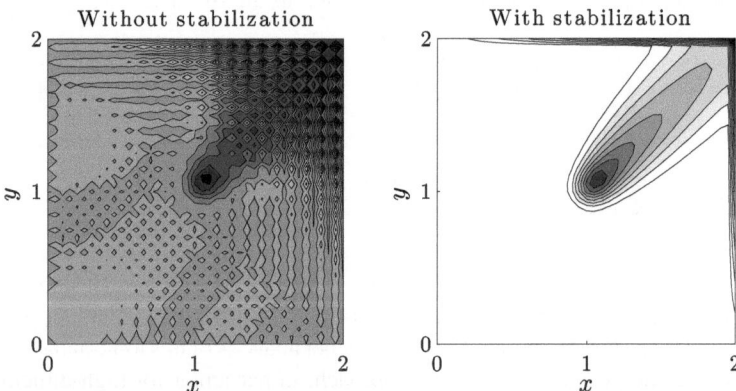

Fig. 6.2 PGD solution of a convection-diffusion problem with (*right*) and without (*left*) convective stabilization

6.3 Discrete Formulation of the PGD

In preparation of our discussion of the last advanced topic, namely the residual minimization strategy, it is useful to present the discrete formulation of the PGD.

Let us for example consider the transient diffusion equation

$$\frac{\partial u}{\partial t} - \Delta u = 0, \tag{6.38}$$

defined in $\Omega \times \Omega_t$ with $\mathbf{x} \in \Omega \subset \mathbb{R}^3$ and $t \in \Omega_t \subset \mathbb{R}^+$.

At enrichment step n of the PGD strategy, the approximate solution is sought in the form

$$u^n(x, y) = \sum_{i=1}^{n-1} X_i(\mathbf{x}) \cdot T_i(t) + R(\mathbf{x}) \cdot S(t), \tag{6.39}$$

where all terms are known with the exception of $R(\mathbf{x})$ and $S(t)$. Again for the sake of simplicity we use $R(\mathbf{x}) \cdot S(t)$ instead of $X_n(\mathbf{x}) \cdot T_n(t)$.

The corresponding weighted residual formulation of the problem reads

$$\int_\Omega \int_{\Omega_t} u^* \cdot \left(\frac{\partial u}{\partial t} - \Delta u \right) \cdot d\mathbf{x} \cdot dt = 0, \tag{6.40}$$

for all suitable test functions u^*. Introducing (6.39) into (6.40) we obtain

$$\int_\Omega \int_{\Omega_t} u^* \cdot \left(R \cdot \frac{\partial S}{\partial t} - \Delta R \cdot S \right) d\mathbf{x} \cdot dt$$

$$= -\int_\Omega \int_{\Omega_t} u^* \cdot \left(\sum_{i=1}^{n-1} \left(X_i \cdot \frac{\partial T_i}{\partial t} - \Delta X_i \cdot T_i \right) \right) d\mathbf{x} \cdot dt. \tag{6.41}$$

This defines a non-linear problem for the unknown functions $R(\mathbf{x})$ and $S(t)$. With the choice $u^* = R^* \cdot S^*$, and in view of the separated expression of all terms involved in the above integrals, we can re-write (6.41) as follows

$$\left(\int_\Omega R^* \cdot R \, d\mathbf{x} \right) \cdot \left(\int_{\Omega_t} S^* \cdot \frac{\partial S}{\partial t} \, dt \right) - \left(\int_\Omega R^* \cdot \Delta R \, d\mathbf{x} \right) \cdot \left(\int_{\Omega_t} S^* \cdot S \, dt \right)$$

$$= -\sum_{i=1}^{n-1} \left(\left(\int_\Omega R^* \cdot X_i \, d\mathbf{x} \right) \cdot \left(\int_{\Omega_t} S^* \cdot \frac{\partial T_i}{\partial t} \, dt \right) \right.$$

$$\left. - \left(\int_\Omega R^* \cdot \Delta X_i \, d\mathbf{x} \right) \cdot \left(\int_{\Omega_t} S^* \cdot T_i \, dt \right) \right). \tag{6.42}$$

Now, each integral term in (6.42) is discretized by using finite elements. Upwinding is considered in the terms involving first derivatives and integration by parts in those involving second derivatives, as discussed in the Appendix. If we denote by $\mathbf{R}$, $\mathbf{S}$, $\mathbf{X}_i$, $\mathbf{T}_i$, $\mathbf{R}^*$ and $\mathbf{S}^*$ the vectors containing the nodal values of the functions R, S, X_i, T_i, R^* and S^*, the discrete form of (6.42) is then given by

$$\left(\mathbf{R}^{*T} \cdot \mathbf{M} \cdot \mathbf{R}\right) \cdot \left(\mathbf{S}^{*T} \cdot \mathbf{C} \cdot \mathbf{S}\right) - \left(\mathbf{R}^{*T} \cdot \mathbf{K} \cdot \mathbf{R}\right) \cdot \left(\mathbf{S}^{*T} \cdot \mathbf{M} \cdot \mathbf{S}\right)$$

$$= -\sum_{i=1}^{n-1} \left(\left(\mathbf{R}^{*T} \cdot \mathbf{M} \cdot \mathbf{X}_i\right) \cdot \left(\mathbf{S}^{*T} \cdot \mathbf{C} \cdot \mathbf{T}_i\right) - \left(\mathbf{R}^{*T} \cdot \mathbf{K} \cdot \mathbf{X}_i\right) \cdot \left(\mathbf{S}^{*T} \cdot \mathbf{M} \cdot \mathbf{T}_i\right)\right).$$

$$(6.43)$$

This is the discrete formulation of the PGD that we shall use in the final section of this chapter.

6.3.1 Example

We now illustrate the construction of matrices $\mathbf{M}$, $\mathbf{K}$ et $\mathbf{C}$ for the particular case of the one-dimensional transient diffusion equation:

- The matrix $\mathbf{M}$ originates from the first term in (6.42),

$$\int_\Omega R^*(x) \cdot R(x)\, dx. \tag{6.44}$$

The function $R(x)$ can be approximated by means of a piecewise linear interpolation from its values $R_i = R(x_i)$ at nodes x_i, $i = 1, \ldots, M$, according to

$$R(x) = \sum_{i=1}^{M} R(x_i) \cdot N_i(x) = \sum_{i=1}^{M} R_i \cdot N_i(x), \tag{6.45}$$

where for the sake of simplicity we assume the nodes distributed uniformly in Ω such that the distance between two consecutive nodes is h. Thus the interpolation functions $N_i(x)$ read for $1 < i < M$:

$$N_i(x) = \begin{cases} \frac{x - x_{i-1}}{h} & x \in [x_{i-1}, x_i] \\ \frac{x_{i+1} - x}{h} & x \in [x_i, x_{i+1}] \\ 0 & elsewhere \end{cases} . \tag{6.46}$$

Moreover, the functions $N_1(x)$ and $N_M(x)$ are given by

$$N_1(x) = \frac{x_2 - x}{h}, \quad x \in [x_1, x_2], \tag{6.47}$$

and

$$N_M(x) = \frac{x - x_{M-1}}{h}, \quad x \in [x_{M-1}, x_M].$$ (6.48)

They both vanish elsewhere.

Thus, the integral (6.44) is given by

$$\int_\Omega R^*(x) \cdot R(x) \, dx = \int_\Omega \left(\sum_{i=1}^M R_i^* \cdot N_i(x) \right) \cdot \left(\sum_{j=1}^M R_j \cdot N_j(x) \right) dx,$$ (6.49)

which can be written in the matrix form

$$\int_\Omega R^*(x) \cdot R(x) \, dx = \mathbf{R}^{*T} \cdot \mathbf{M} \cdot \mathbf{R}.$$ (6.50)

Here, we have $\mathbf{M}_{ij} = \int_\Omega N_i(x) \cdot N_j(x) \, dx$, so that $\mathbf{M}_{ii} = \frac{2h}{3}$ for $1 < i < M$, $\mathbf{M}_{11} = \mathbf{M}_{MM} = \frac{h}{3}$, $\mathbf{M}_{ij} = \frac{h}{6}$ if $|i - j| = 1$ and $\mathbf{M}_{ij} = 0$ if $|i - j| > 1$.

- We now consider terms involving the second derivative of the space coordinate, as for example

$$\int_\Omega R^* \cdot \frac{d^2 R}{dx^2} \, dx.$$ (6.51)

Integration by parts yields

$$\int_\Omega R^* \cdot \frac{d^2 R}{dx^2} \, dx = -\int_\Omega \frac{d R^*}{dx} \cdot \frac{d R}{dx} \, dx,$$ (6.52)

where the boundary integral vanishes if Dirichlet boundary conditions are specified (see Appendix).

By considering the approximation derivative, we obtain

$$\frac{d R(x)}{dx} = \sum_{i=1}^M R(x_i) \cdot \frac{d N_i(x)}{dx} = \sum_{i=1}^M R_i \cdot \frac{d N_i(x)}{dx}.$$ (6.53)

Here, the derivatives $\frac{d N_i(x)}{dx}$ of the interpolation functions read for $1 < i < M$:

$$\frac{d N_i(x)}{dx} = \begin{cases} \frac{1}{h} & x \in [x_{i-1}, x_i] \\ -\frac{1}{h} & x \in [x_i, x_{i+1}] \\ 0 & elsewhere \end{cases}.$$ (6.54)

Moreover, the derivatives $\frac{d N_1(x)}{dx}$ and $\frac{d N_M(x)}{dx}$ are given by

$$\frac{dN_1(x)}{dx} = -\frac{1}{h}, \quad x \in [x_1, x_2], \tag{6.55}$$

and

$$\frac{dN_M(x)}{dx} = \frac{1}{h}, \quad x \in [x_{M-1}, x_M]. \tag{6.56}$$

They both vanish elsewhere.

Thus, the integral (6.52) becomes

$$-\int_\Omega \frac{dR^*}{dx} \cdot \frac{dR}{dx} dx = -\int_\Omega \left(\sum_{i=1}^M R_i^* \cdot \frac{dN_i(x)}{dx} \right) \cdot \left(\sum_{j=1}^M R_j \cdot \frac{dN_j(x)}{dx} \right) dx, \tag{6.57}$$

which can be written in the matrix form

$$-\int_\Omega \frac{dR^*}{dx} \cdot \frac{dR}{dx} dx = \mathbf{R}^{*T} \cdot \mathbf{K} \cdot \mathbf{R}. \tag{6.58}$$

Here, we have $\mathbf{K}_{ij} = \int_\Omega \frac{dN_i(x)}{dx} \cdot \frac{dN_j(x)}{dx} dx$, so that $\mathbf{K}_{ii} = -\frac{2}{h^2}$ for $1 < i < M$, $\mathbf{K}_{11} = \mathbf{K}_{MM} = -\frac{1}{h^2}$, $\mathbf{K}_{ij} = \frac{1}{h^2}$ if $|i - j| = 1$ and $\mathbf{K}_{ij} = 0$ if $|i - j| > 1$.

- Finally, we consider a term involving first-order time derivatives, as for example

$$\int_{\Omega_t} S^* \cdot \frac{dS}{dt} dt. \tag{6.59}$$

We approximate the function $S(t)$ by means of piecewise linear functions $N_i(t)$ associated to a uniform discretization of the time interval with a constant time step Δt,

$$S(t) = \sum_{i=1}^M S(t_i) \cdot N_i(t) = \sum_{i=1}^M S_i \cdot N_i(t). \tag{6.60}$$

The interpolating functions are given by

$$N_i(t) = \begin{cases} \frac{t - t_{i-1}}{\Delta t} & t \in [t_{i-1}, t_i] \\ \frac{t_{i+1} - t}{\Delta t} & t \in [t_i, t_{i+1}] \\ 0 & elsewhere \end{cases}. \tag{6.61}$$

Moreover, the functions $N_1(t)$ and $N_M(t)$ read

$$N_1(t) = \frac{t_2 - t}{\Delta t}, \quad t \in [t_1, t_2], \tag{6.62}$$

and

$$N_M(t) = \frac{t - t_{M-1}}{\Delta t}, \quad t \in [t_{M-1}, t_M]. \tag{6.63}$$

They both vanish elsewhere.

The approximation of $\frac{dS(t)}{dt}$ is thus given by

$$\frac{dS(t)}{dt} = \sum_{i=1}^{M} S(t_i) \cdot D_i(t) = \sum_{i=1}^{M} S_i \cdot D_i(t), \qquad (6.64)$$

where the functions $D_i(t)$ read

$$D_i(t) = \begin{cases} \frac{1}{\Delta t} & x \in [x_{i-1}, x_i] \\ 0 & elsewhere \end{cases}. \qquad (6.65)$$

Thus, the integral (6.59) becomes

$$\int_{\Omega_t} S^* \cdot \frac{dS(t)}{dt} \, dt = \int_{\Omega_t} \left(\sum_{i=1}^{M} S_i^* \cdot N_i(t) \right) \cdot \left(\sum_{j=1}^{M} S_j \cdot D_j(t) \right) dt, \qquad (6.66)$$

which can be written in the matrix form

$$\int_{\Omega_t} S^* \cdot \frac{dS(t)}{dt} \, dt = \mathbf{S}^{*T} \cdot \mathbf{C} \cdot \mathbf{S}. \qquad (6.67)$$

with $\mathbf{C}_{ij} = \int_{\Omega_t} N_i(t) \cdot D_j(t) \, dt$.

For more complex situations (2D or 3D space problems), the interested reader can refer to the finite element primer given in the Appendix.

6.3.2 Alternating Direction Scheme

The discrete form of the alternating direction iterative scheme follows directly:

- Compute $\mathbf{R}$ by assuming $\mathbf{S}^* = \mathbf{S}$: we get from (6.43)

$$\mathbf{R}^{*T} \cdot \mathbf{A} \cdot \mathbf{R} = \mathbf{R}^{*T} \cdot \mathbf{F}, \qquad (6.68)$$

which, in view of the arbitrariness of $\mathbf{R}^*$, yields the linear algebraic problem

$$\mathbf{A} \cdot \mathbf{R} = \mathbf{F}. \qquad (6.69)$$

- Compute $\mathbf{S}$ by assuming $\mathbf{R}^* = \mathbf{R}$: we get from (6.43)

$$\mathbf{S}^{*T} \cdot \mathbf{B} \cdot \mathbf{S} = \mathbf{S}^{*T} \cdot \mathbf{G}, \qquad (6.70)$$

which, in view of the arbitrariness of $\mathbf{S}^*$, yields the linear algebraic problem

$$\mathbf{B} \cdot \mathbf{S} = \mathbf{G}. \tag{6.71}$$

We leave it as an exercise to write down the detailed expressions for the discrete terms $\mathbf{A}$, $\mathbf{F}$, $\mathbf{B}$ and $\mathbf{G}$.

6.4 Residual Minimization for Non-Symmetric Operators

The standard PGD strategy detailed in the previous chapters can fail to converge, or converges too slowly, when applied to non-symmetric differential operators. An alternative strategy based on the idea of residual minimization can be used that converges in most situations, even if it produces non-optimal separated representations containing more terms than strictly needed, or, sometimes, converges slowly due to degradation in conditioning.

Consider a linear system of equations

$$\mathbf{A} \cdot \mathbf{X} = \mathbf{F}, \tag{6.72}$$

whose residual is $\mathbf{A} \cdot \mathbf{X} - \mathbf{F}$. The solution $\mathbf{X}$ minimizes the residual norm $\|\mathbf{A} \cdot \mathbf{X} - \mathbf{F}\|^2$. The extremum condition implies

$$\left(\mathbf{A}^T \cdot \mathbf{A} \right) \cdot \mathbf{X} = \mathbf{A}^T \cdot \mathbf{F}. \tag{6.73}$$

In view of the symmetry of $\mathbf{A}^T \cdot \mathbf{A}$, this problem can be solved by using strategies adapted to symmetric algebraic systems.

We roughly follow this idea in the context of the PGD. In the last section, we arrived for problem (6.38) at

$$\left(\mathbf{R}^{*T} \cdot \mathbf{M} \cdot \mathbf{R} \right) \cdot \left(\mathbf{S}^{*T} \cdot \mathbf{C} \cdot \mathbf{S} \right) - \left(\mathbf{R}^{*T} \cdot \mathbf{K} \cdot \mathbf{R} \right) \cdot \left(\mathbf{S}^{*T} \cdot \mathbf{M} \cdot \mathbf{S} \right) =$$

$$- \sum_{i=1}^{n-1} \left(\left(\mathbf{R}^{*T} \cdot \mathbf{M} \cdot \mathbf{X}_i \right) \cdot \left(\mathbf{S}^{*T} \cdot \mathbf{C} \cdot \mathbf{T}_i \right) - \left(\mathbf{R}^{*T} \cdot \mathbf{K} \cdot \mathbf{X}_i \right) \cdot \left(\mathbf{S}^{*T} \cdot \mathbf{M} \cdot \mathbf{T}_i \right) \right). \tag{6.74}$$

Again for the sake of simplicity and without loss of generality, let us assume that we are at the second enrichment step ($n = 2$). Thus, (6.74) reduces to

$$\left(\mathbf{R}^{*T} \cdot \mathbf{M} \cdot \mathbf{R} \right) \cdot \left(\mathbf{S}^{*T} \cdot \mathbf{C} \cdot \mathbf{S} \right) - \left(\mathbf{R}^{*T} \cdot \mathbf{K} \cdot \mathbf{R} \right) \cdot \left(\mathbf{S}^{*T} \cdot \mathbf{M} \cdot \mathbf{S} \right)$$

$$= - \left(\mathbf{R}^{*T} \cdot \mathbf{M} \cdot \mathbf{X}_1 \right) \cdot \left(\mathbf{S}^{*T} \cdot \mathbf{C} \cdot \mathbf{T}_1 \right) + \left(\mathbf{R}^{*T} \cdot \mathbf{K} \cdot \mathbf{X}_1 \right) \cdot \left(\mathbf{S}^{*T} \cdot \mathbf{M} \cdot \mathbf{T}_1 \right). \tag{6.75}$$

The residual $\mathbf{R}e$ can be written as

$$\mathbf{R}e = (\mathbf{M} \cdot \mathbf{R}) \cdot (\mathbf{C} \cdot \mathbf{S})^T - (\mathbf{K} \cdot \mathbf{R}) \cdot (\mathbf{M} \cdot \mathbf{S})^T$$

$$+ (\mathbf{M} \cdot \mathbf{X}_1) \cdot (\mathbf{C} \cdot \mathbf{T}_1)^T - (\mathbf{K} \cdot \mathbf{X}_1) \cdot (\mathbf{M} \cdot \mathbf{T}_1)^T , \qquad (6.76)$$

and the square of its norm is given by

$$\|\mathbf{R}e\|^2 = (\mathbf{M}{\cdot}\mathbf{R})^2{\cdot}(\mathbf{C}{\cdot}\mathbf{S})^2 + (\mathbf{K}{\cdot}\mathbf{R})^2{\cdot}(\mathbf{M}{\cdot}\mathbf{S})^2 + (\mathbf{M}{\cdot}\mathbf{X}_1)^2{\cdot}(\mathbf{C}{\cdot}\mathbf{T}_1)^2 + (\mathbf{K}{\cdot}\mathbf{X}_1)^2{\cdot}(\mathbf{M}{\cdot}\mathbf{T}_1)^2$$

$$-2 \cdot ((\mathbf{M} \cdot \mathbf{R}) \cdot (\mathbf{K} \cdot \mathbf{R})) \cdot ((\mathbf{C} \cdot \mathbf{S}) \cdot (\mathbf{M} \cdot \mathbf{S}))$$

$$+2 \cdot ((\mathbf{M} \cdot \mathbf{R}) \cdot (\mathbf{M} \cdot \mathbf{X}_1)) \cdot ((\mathbf{C} \cdot \mathbf{S}) \cdot (\mathbf{C} \cdot \mathbf{T}_1))$$

$$-2 \cdot ((\mathbf{M} \cdot \mathbf{R}) \cdot (\mathbf{K} \cdot \mathbf{X}_1)) \cdot ((\mathbf{C} \cdot \mathbf{S}) \cdot (\mathbf{M} \cdot \mathbf{T}_1))$$

$$-2 \cdot ((\mathbf{K} \cdot \mathbf{R}) \cdot (\mathbf{M} \cdot \mathbf{X}_1)) \cdot ((\mathbf{M} \cdot \mathbf{S}) \cdot (\mathbf{C} \cdot \mathbf{T}_1))$$

$$+2 \cdot ((\mathbf{K} \cdot \mathbf{R}) \cdot (\mathbf{K} \cdot \mathbf{X}_1)) \cdot ((\mathbf{M} \cdot \mathbf{S}) \cdot (\mathbf{M} \cdot \mathbf{T}_1))$$

$$- 2 \cdot ((\mathbf{M} \cdot \mathbf{X}_1) \cdot (\mathbf{K} \cdot \mathbf{X}_1)) \cdot ((\mathbf{C} \cdot \mathbf{T}_1) \cdot (\mathbf{M} \cdot \mathbf{T}_1)) . \qquad (6.77)$$

The unknown $\mathbf{R}$ and $\mathbf{S}$ will be obtained by enforcing residual minimization, which implies

$$\begin{cases} \dfrac{\partial \|\mathbf{R}e\|^2}{\partial \mathbf{R}} = \mathbf{0} \\[2mm] \dfrac{\partial \|\mathbf{R}e\|^2}{\partial \mathbf{S}} = \mathbf{0} \end{cases} . \qquad (6.78)$$

Let us now consider the detailed expression of the three types of derivatives with respect to $\mathbf{R}$ involved in (6.77).

- Symmetric quadratic terms, as for example $(\mathbf{M} \cdot \mathbf{R})^2$, whose derivative can be computed by considering

$$(\mathbf{R}+\delta\mathbf{R})^T \cdot \mathbf{M}^T \cdot \mathbf{M} \cdot (\mathbf{R}+\delta\mathbf{R}) - \mathbf{R}^T \cdot \mathbf{M}^T \cdot \mathbf{M} \cdot \mathbf{R} \approx \left(\frac{\partial (\mathbf{M} \cdot \mathbf{R})^2}{\partial \mathbf{R}} \right)^T \cdot \delta\mathbf{R}. \quad (6.79)$$

Neglecting second-order terms in $\delta\mathbf{R}$, we obtain

$$\delta\mathbf{R}^T \cdot \mathbf{M}^T \cdot \mathbf{M} \cdot \mathbf{R} + \mathbf{R}^T \cdot \mathbf{M}^T \cdot \mathbf{M} \cdot \delta\mathbf{R} \approx \left(\frac{\partial (\mathbf{M} \cdot \mathbf{R})^2}{\partial \mathbf{R}} \right)^T \cdot \delta\mathbf{R}, \qquad (6.80)$$

or

$$\mathbf{R}^T \cdot \mathbf{M}^T \cdot \mathbf{M} \cdot \delta\mathbf{R} + \mathbf{R}^T \cdot \mathbf{M}^T \cdot \mathbf{M} \cdot \delta\mathbf{R} \approx \left(\frac{\partial(\mathbf{M} \cdot \mathbf{R})^2}{\partial\mathbf{R}}\right)^T \cdot \delta\mathbf{R}, \quad (6.81)$$

from which we find

$$\left(\frac{\partial(\mathbf{M} \cdot \mathbf{R})^2}{\partial\mathbf{R}}\right)^T = 2 \cdot \mathbf{R}^T \cdot \mathbf{M}^T \cdot \mathbf{M} \rightarrow \frac{\partial(\mathbf{M} \cdot \mathbf{R})^2}{\partial\mathbf{R}} = 2 \cdot \mathbf{M}^T \cdot \mathbf{M} \cdot \mathbf{R}. \quad (6.82)$$

Notice the symmetry of $\mathbf{M}^T \cdot \mathbf{M}$.

- Non-symmetric quadratic terms, as for example $(\mathbf{M} \cdot \mathbf{R}) \cdot (\mathbf{K} \cdot \mathbf{R})$, whose derivative can be computed by considering

$$(\mathbf{R}+\delta\mathbf{R})^T \cdot \mathbf{M}^T \cdot \mathbf{K} \cdot (\mathbf{R}+\delta\mathbf{R}) - \mathbf{R}^T \cdot \mathbf{M}^T \cdot \mathbf{K} \cdot \mathbf{R} \approx \left(\frac{\partial((\mathbf{M} \cdot \mathbf{R}) \cdot (\mathbf{K} \cdot \mathbf{R}))}{\partial\mathbf{R}}\right)^T \cdot \delta\mathbf{R}. \quad (6.83)$$

Neglecting second-order terms in $\delta\mathbf{R}$, we get

$$\delta\mathbf{R}^T \cdot \mathbf{M}^T \cdot \mathbf{K} \cdot \mathbf{R} + \mathbf{R}^T \cdot \mathbf{M}^T \cdot \mathbf{K} \cdot \delta\mathbf{R} \approx \left(\frac{\partial((\mathbf{M} \cdot \mathbf{R}) \cdot (\mathbf{K} \cdot \mathbf{R}))}{\partial\mathbf{R}}\right)^T \cdot \delta\mathbf{R}, \quad (6.84)$$

or

$$\mathbf{R}^T \cdot \mathbf{K}^T \cdot \mathbf{M} \cdot \delta\mathbf{R} + \mathbf{R}^T \cdot \mathbf{M}^T \cdot \mathbf{K} \cdot \delta\mathbf{R} \approx \left(\frac{\partial((\mathbf{M} \cdot \mathbf{R}) \cdot (\mathbf{K} \cdot \mathbf{R}))}{\partial\mathbf{R}}\right)^T \cdot \delta\mathbf{R}, \quad (6.85)$$

and thus

$$\left(\frac{\partial((\mathbf{M} \cdot \mathbf{R}) \cdot (\mathbf{K} \cdot \mathbf{R}))}{\partial\mathbf{R}}\right)^T = \mathbf{R}^T \cdot \left(\mathbf{M}^T \cdot \mathbf{K} + \mathbf{K}^T \cdot \mathbf{M}\right) \rightarrow$$

$$\frac{\partial((\mathbf{M} \cdot \mathbf{R}) \cdot (\mathbf{K} \cdot \mathbf{R}))}{\partial\mathbf{R}} = \left(\mathbf{M}^T \cdot \mathbf{K} + \mathbf{K}^T \cdot \mathbf{M}\right) \cdot \mathbf{R}. \quad (6.86)$$

Note again the symmetry of $\mathbf{M}^T \cdot \mathbf{K} + \mathbf{K}^T \cdot \mathbf{M}$.

- Linear terms, as for example $(\mathbf{M} \cdot \mathbf{R}) \cdot (\mathbf{K} \cdot \mathbf{X}_1)$, whose derivative can be computed by considering

$$\mathbf{X}_1^T \cdot \mathbf{K}^T \cdot \mathbf{M} \cdot (\mathbf{R}+\delta\mathbf{R}) - \mathbf{X}_1^T \cdot \mathbf{K}^T \cdot \mathbf{M} \cdot \mathbf{R} \approx \left(\frac{\partial((\mathbf{M} \cdot \mathbf{R}) \cdot (\mathbf{K} \cdot \mathbf{X}_1))}{\partial\mathbf{R}}\right)^T \cdot \delta\mathbf{R}, \quad (6.87)$$

which yields

$$\mathbf{X}_1^T \cdot \mathbf{K}^T \cdot \mathbf{M} \cdot \delta\mathbf{R} \approx \left(\frac{\partial((\mathbf{M} \cdot \mathbf{R}) \cdot (\mathbf{K} \cdot \mathbf{X}_1))}{\partial\mathbf{R}}\right)^T \cdot \delta\mathbf{R}. \quad (6.88)$$

We thus finds that

$$\left(\frac{\partial((\mathbf{M} \cdot \mathbf{R}) \cdot (\mathbf{K} \cdot \mathbf{X}_1))}{\partial \mathbf{R}}\right)^T = \mathbf{X}_1^T \cdot \mathbf{K}^T \cdot \mathbf{M} \rightarrow$$

$$\frac{\partial((\mathbf{M} \cdot \mathbf{R}) \cdot (\mathbf{K} \cdot \mathbf{X}_1))}{\partial \mathbf{R}} = \mathbf{M}^T \cdot \mathbf{K} \cdot \mathbf{X}_1. \tag{6.89}$$

As expected, the derivative of linear terms does not involve the unknown function.

We notice the symmetrization effect of the residual minimization strategy. Similar relationships are obtained for the derivative of the residual with respect to $\mathbf{S}$.

Thus, with $\mathbf{X}_1$ and $\mathbf{T}_1$ known,

- If we assume $\mathbf{S}$ known, the condition

$$\frac{\partial \|\mathbf{R}e\|^2}{\partial \mathbf{R}} = \mathbf{0}, \tag{6.90}$$

results in a symmetric linear algebraic system for computing $\mathbf{R}$.
- Then, with the just computed $\mathbf{R}$, the condition

$$\frac{\partial \|\mathbf{R}e\|^2}{\partial \mathbf{S}} = \mathbf{0}, \tag{6.91}$$

results in a symmetric linear algebraic system for updating $\mathbf{S}$.
- The iteration continues until reaching the convergence of the alternated direction scheme; we then assign $\mathbf{R} \rightarrow \mathbf{X}_2$ and $\mathbf{S} \rightarrow \mathbf{T}_2$, and move to the next enrichment step.

This procedure can be generalized to generic multidimensional models as described in [13].

6.5 Numerical Example

In this final section, we consider a multidimensional parametric problem which combines many of the ingredients described in the previous chapters. The problem is defined by the following partial differential equation for the unknown function u:

$$\mathbf{v} \cdot \nabla u - k \cdot \Delta u = f, \tag{6.92}$$

with $\mathbf{v} = (v_x \ v_y)^T$, $u = u(x, y, v_x, v_y, k)$ and $(x, y, v_x, v_y, k) \in \Omega$. The domain Ω is defined as $\Omega = \Omega_x \times \Omega_y \times \Omega_{v_x} \times \Omega_{v_y} \times \Omega_k$, with

$$\begin{cases} \Omega_x = (0, 2) \\ \Omega_y = (0, 2) \\ \Omega_{v_x} = [0.5, 1] \\ \Omega_{v_y} = [0.5, 1] \\ \Omega_k = [10^{-3}, 10^{-1}] \end{cases} \quad . \tag{6.93}$$

The source term f is defined as

$$f(x, y) = 10 \, e^{-100(x-1)^2} \, e^{-100(y-1)^2}, \tag{6.94}$$

and the boundary conditions are again given by

$$\begin{cases} u(0, y) = u(x, 0) = 0 \\ u(x, 2) = \frac{x}{2} \\ u(2, y) = \frac{y}{2} \end{cases} \quad . \tag{6.95}$$

In this particular high-dimensional problem, the non-symmetry induced by the convection term requires use of the residual minimization technique for the PGD

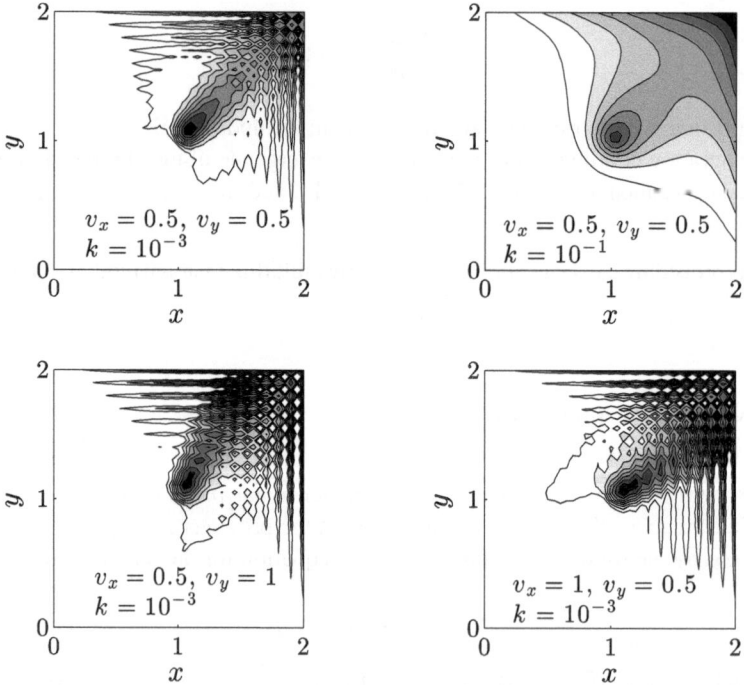

Fig. 6.3 PGD solution of problem (6.92) without convective stabilization and particularized for different values of v_x, v_y and k

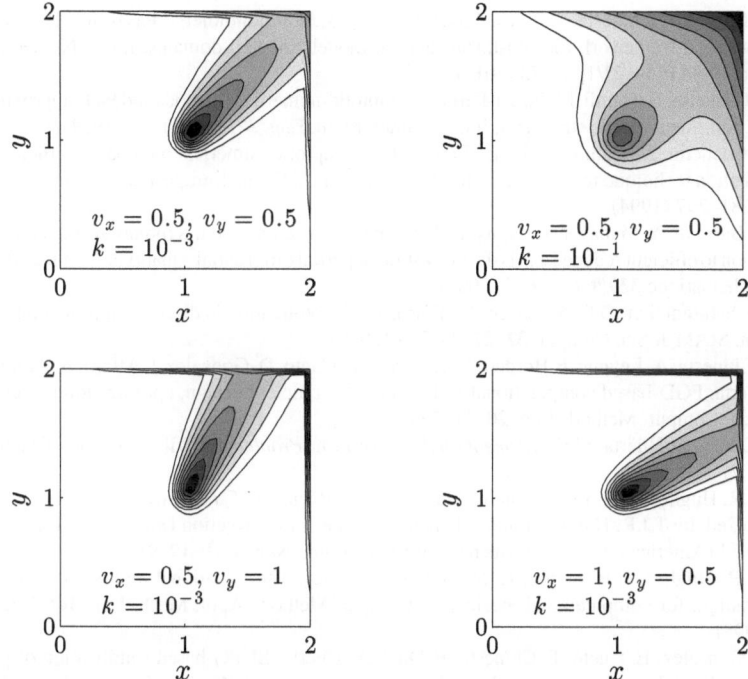

Fig. 6.4 PGD solution of problem (6.92) with convective stabilization and particularized for different values of v_x, v_y and k

strategy to converge. All one-dimensional problems are solved using a finite difference method, with or without convective stabilization.

In Fig. 6.3, we show the computed PGD solution obtained without convective stabilization and particularized for different values of v_x, v_y and k. For low values of the diffusivity, the lack of convective stabilization induces spurious numerical oscillations. Convective stabilization is not necessary, however for higher diffusivities.

In Fig. 6.4 we show the same results but now obtained with convective stabilization. The PGD solution is clearly devoid of spurious oscillations for all values of the diffusivity. Moreover, use of convective stabilization requires a much smaller number of enrichment steps.

References

1. A. Ammar, M. Normandin, F. Daim, D. Gonzalez, E. Cueto, F. Chinesta, Non-incremental strategies based on separated representations: applications in computational rheology. Commun. Math. Sci. **8/3**, 671–695 (2010)
2. E. Pruliere, F. Chinesta, A. Ammar, On the deterministic solution of multidimensional parametric models by using the Proper Generalized Decomposition. Math. Comput. Simul. **81**, 791–810 (2010)

3. A. Leygue, F. Chinesta, M. Beringhier, T.L. Nguyen, J.C. Grandidier, F. Pasavento, B. Schrefler, Towards a framework for non-linear thermal models in shell domains. Int. J. Numer. Meth. Heat Fluid Flow 23/1, 55–73 (2013)

4. B. Cochelin, N. Damil, M. Potier-Ferry, Asymptotic-numerical methods and Pade approximants for non-linear elastic structures. Int. J. Numer. Meth. Eng. **37**, 1187–1213 (1994)

5. B. Cochelin, N. Damil, M. Potier-Ferry, The asymptotic numerical method: an efficient perturbation technique for nonlinear structural mechanics. Revue Europeenne des Elements Finis **3**, 281–297 (1994)

6. M. Barrault, Y. Maday, N.C. Nguyen, A.T. Patera, An "empirical interpolation" method: application to efficient reduced-basis discretization of partial differential equations. Comptes Rendus Mathematique **339/9**, 667–672 (2004)

7. S. Chaturantabut, D.C. Sorensen, Nonlinear model reduction via discrete empirical interpolation. SIAM J. Sci. Comput. **32**, 2737–2764 (2010)

8. F. Chinesta, A. Leygue, F. Bordeu, J.V. Aguado, E. Cueto, D. Gonzalez, I. Alfaro, A. Ammar, A. Huerta, PGD-based computational vademecum for efficient design, optimization and control. Arch. Comput. Methods Eng. **20**, 31–59 (2013)

9. J. Donea, A. Huerta, *Finite Element Methods for Flow Problems* (J Wiley and Sons, Chichester, 2002)

10. T.J.R. Hughes, A N. Brooks, in *A Multidimensional Upwind Scheme with no Crosswind Difusion*, ed. by T.J.R. Hughes. Finite Element Methods for Convection Dominated Flows. AMD, vol 34 (American Society of Mechanical Engineering, New York, 1979)

11. T.J R. Hughes, G.R. Feijóo, L. Mazzei, J-B. Quincy, The variational multiscale method—a paradigm for computational mechanics. Comput. Methods Appl. Mech. Eng. **166/1-2**, 3–24 (1998)

12. D. Gonzalez, E. Cueto, F. Chinesta, P. Diez, A. Huerta, SUPG-based stabilization of proper generalized decompositions for high-dimensional advection-diffusion equations. Int. J. Numer. Meth. Eng. **94/13**, 1216–1232 (2013)

13. F. Chinesta, A. Ammar, E. Cueto, Recent advances and new challenges in the use of the Proper Generalized Decomposition for solving multidimensional models. Arch. Comput. Methods Eng. **17/4**, 327–350 (2010)

Appendix A
Standard Discretization Techniques

Abstract In this appendix, we summarize some classical discretization techniques able to solve initial value and boundary value problems. In particular, we briefly revisit the main ideas behind finite difference and finite element methods.

The PGD methodology described throughout this book leads to the numerical solution of decoupled low-dimensional problems governed by ordinary or partial differential equations. In this appendix, we briefly summarize the most popular methods for solving such problems, namely finite difference and finite element techniques. For more detail, the reader is invited to consult the prolific literature on these classical subjects.

A.1 First-Order Ordinary Differential Equations

We consider in this section a typical first-order ordinary differential equation for the unknown function $u(t)$,

$$\alpha \cdot \frac{du}{dt} + \beta \cdot u = f(t). \tag{A.1}$$

Here, α and β are constant coefficients, and $f(t)$ is a given forcing term. The problem is defined in the time interval $\mathcal{I} = (0, T]$, with the initial condition $u(t = 0) = u_0$.

The simplest discretization method is known as the Euler explicit finite difference scheme. We consider a time step δt and wish to compute a numerical approximation u_i of $u(t_i)$ at discrete values of time $t_i = i \cdot \delta t$. The simplest approach consists in approximating the time derivative at time t_i by a first-order, backward finite difference:

$$\frac{du}{dt}\Big|_i \approx \frac{u_i - u_{i-1}}{\delta t}. \tag{A.2}$$

Equation (A.1) evaluated at time t_i then reads

F. Chinesta et al., *The Proper Generalized Decomposition for Advanced Numerical Simulations*, SpringerBriefs in Applied Sciences and Technology, DOI: 10.1007/978-3-319-02865-1, © The Author(s) 2014

$$\alpha \cdot \frac{u_i - u_{i-1}}{\delta t} + \beta \cdot u_i = f(t_i), \tag{A.3}$$

which yields a simple recurrence for the unknown u_i,

$$(\alpha + \beta \cdot \delta t) \cdot u_i = \delta t \cdot f(t_i) + \alpha \cdot u_{i-1}. \tag{A.4}$$

Thus, with u_0 given (the initial condition), we obtain via (A.4) with $i = 1$ the approximate solution u_1 at time $t_1 = \delta t$, and then u_2 from u_1 and so on, until reaching the final time T. This simple incremental technique does not require the solution of an algebraic system. The procedure is thus extremely simple to implement, cheap and fast from the computational point of view. Note however that the time step δt must be small enough to ensure numerical stability and solution accuracy.

A.2 Partial Differential Equations: Finite Differences

We consider the Poisson equation for the unknown function $u(\mathbf{x})$,

$$\Delta u(\mathbf{x}) = f(\mathbf{x}), \tag{A.5}$$

defined in a two-dimensional rectangular domain $\Omega = \Omega_x \times \Omega_y = (0, L) \times (0, H)$, with $\mathbf{x} = (x, y) \in \Omega$. The forcing term $f(\mathbf{x})$ is given. The boundary Γ of Ω is composed of two disjoint parts Γ_D and Γ_N, where Dirichlet and Neumann boundary conditions apply respectively:

$$\begin{cases} u(\mathbf{x} \in \Gamma_D) = u_g \\ \nabla u \cdot \mathbf{n}|_{\mathbf{x} \in \Gamma_N} = q_g \end{cases}. \tag{A.6}$$

Here, $\mathbf{n}$ is the unit outward normal vector on Γ_N, and u_g, q_g are given functions specified on Γ_D and Γ_N, respectively.

With a finite difference method, we discretize the domain into a grid composed of $N_x \times N_y$ nodes. The coordinates (x_i, y_j) of a generic node ($i = 1, \ldots, N_x$; $j = 1, \ldots, N_y$) are given by

$$\begin{cases} x_i = (i - 1) \cdot h_x \\ y_j = (j - 1) \cdot h_y \end{cases}, \tag{A.7}$$

where the grid spacings $h_x = \frac{L}{N_x - 1}$ and $h_y = \frac{H}{N_y - 1}$ are specified by the user.

In order to compute a numerical approximation $u_{i,j}$ of $u(x_i, y_j)$, the governing Eq. (A.5) is enforced at each internal node (x_i, y_j), i.e. for $1 < i < N_x$ and $1 < j < N_y$, whereas the boundary conditions are enforced at nodes located on the domain boundary Γ.

Finite difference formulae are used to approximate the derivatives appearing in (A.5). For example, the first-order derivative at (x_i, y_j) can be approximated for $1 < i < N_x$ and $1 < j < N_y$ in different ways such as

$$
\begin{cases}
\frac{\partial u}{\partial x}\big|_{i,j} \approx \frac{u_{i,j}-u_{i-1,j}}{h_x} \\[2ex]
\frac{\partial u}{\partial x}\big|_{i,j} \approx \frac{u_{i+1,j}-u_{i,j}}{h_x} \\[2ex]
\frac{\partial u}{\partial x}\big|_{i,j} \approx \frac{u_{i+1,j}-u_{i-1,j}}{2 \cdot h_x}
\end{cases}
\tag{A.8}
$$

Note that the third approximation enjoys second-order accuracy and is the best of the three listed above. Now, since the second derivative is the first derivative applied on the first derivative, we have for example

$$
\frac{\partial^2 u}{\partial x^2}\big|_{i,j} \approx \frac{\frac{\partial u}{\partial x}\big|_{i,j} - \frac{\partial u}{\partial x}\big|_{i-1,j}}{h_x} \approx \frac{u_{i+1,j} - 2 \cdot u_{i,j} + u_{i-1,j}}{h_x^2}.
\tag{A.9}
$$

The governing Eq. (A.5) enforced at each internal node thus reads

$$
\frac{u_{i+1,j} - 2 \cdot u_{i,j} + u_{i-1,j}}{h_x^2} + \frac{u_{i,j+1} - 2 \cdot u_{i,j} + u_{i,j-1}}{h_y^2} = f(x_i, y_j),
\tag{A.10}
$$

for $1 < i < N_x$ and $1 < j < N_y$.

For the nodes located on the boundary, i.e. $i = 1, i = N_x, j = 1$ and $j = N_y$, we specify the boundary conditions. For $(x_i, y_j) \in \Gamma_D$, we write

$$
u_{i,j} = u_g(x_i, y_j),
\tag{A.11}
$$

while for $(x_i, y_j) \in \Gamma_N$, we prescribe

$$
\begin{cases}
\frac{u_{i+1,j}-u_{i,j}}{h_x} = q_g(x_i, y_j) & \text{if } i = 0 \\[2ex]
\frac{u_{i,j}-u_{i-1,j}}{h_x} = q_g(x_i, y_j) & \text{if } i = N_x \\[2ex]
\frac{u_{i,j+1}-u_{i,j}}{h_y} = q_g(x_i, y_j) & \text{if } j = 0 \\[2ex]
\frac{u_{i,j}-u_{i,j-1}}{h_y} = q_g(x_i, y_j) & \text{if } j = N_y
\end{cases}
\tag{A.12}
$$

We have thus obtained a system of $N_x \cdot N_y$ linear algebraic equations for the unknowns $u_{i,j}$.

A.3 Partial Differential Equations: Finite Elements

We consider a two-dimensional domain $\Omega \subset \mathcal{R}^2$ with an arbitrary geometry, wherein a simple heat conduction problem is defined:

$$\nabla \cdot (\mathbf{D} \cdot \nabla u(\mathbf{x})) + f(\mathbf{x}) = 0. \tag{A.13}$$

Here, $\mathbf{x} = (x, y) \in \Omega$ and $\mathbf{D}$ is the conductivity tensor.

The boundary of Ω, $\Gamma \equiv \partial\Omega$, is composed of two disjoint parts Γ_D and Γ_N, where Dirichlet and Neumann boundary conditions apply respectively:

$$\begin{cases} u(\mathbf{x} \in \Gamma_D) = u_g \\ -(\mathbf{D} \cdot \nabla u) \cdot \mathbf{n}|_{\mathbf{x} \in \Gamma_N} = q_g \end{cases}, \tag{A.14}$$

where $\mathbf{n}$ is the unit outward normal vector on Γ_N. We specify homogeneous boundary conditions, that is $u_g = 0$ and $q_g = 0$.

The weighted residual form of (A.13) reads: Find $u(\mathbf{x})$ regular enough and verifying the Dirichlet boundary conditions, $u(\mathbf{x} \in \Gamma_D) = u_g = 0$, such that

$$\int_\Omega u^* \left(\nabla \cdot (\mathbf{D} \cdot \nabla u(\mathbf{x})) + f \right) d\mathbf{x} = 0, \tag{A.15}$$

$\forall u^*$ in an appropriate functional space.

The previous integral form needs $\mathcal{C}^1$ continuity of the approximation of $u(\mathbf{x})$ because of the second derivatives that it involves. To avoid this difficulty, integration by parts is commonly performed to arrive at the associated variational form of (A.13)

$$\int_\Omega (\nabla u^*)^T \cdot \mathbf{D} \cdot \nabla u \, d\mathbf{x} = \int_\Omega u^* \cdot f \, d\mathbf{x}, \tag{A.16}$$

$\forall u^*$ in an appropriate functional space and verifying $u^*(\mathbf{x} \in \Gamma_D) = 0$.

Now, in order to address general complex geometries, we associate a mesh to Ω, that is a decomposition of Ω in a series of elements Ω^e, $e = 1, \ldots, E$, such that $\Omega^i \cap \Omega^j = \emptyset$ and $\Omega = \bigcup_{e=1}^{e=E} \Omega^e$.

The weak form (A.16) can be rewritten as

$$\sum_{e=1}^{e=E} \int_{\Omega^e} (\nabla u^{e*})^T \cdot \mathbf{D} \cdot \nabla u^e \, d\mathbf{x} = \sum_{e=1}^{e=E} \int_{\Omega^e} u^{e*} \cdot f \, d\mathbf{x}, \tag{A.17}$$

where $u^e(\mathbf{x})$ is the restriction of $u(\mathbf{x})$ to Ω^e.

The above expression implies to approximate the unknown field $u(\mathbf{x})$ in each element Ω^e. To simplify the computational procedure, it is preferable to define the approximation in a unique reference element Ω^r and perform a geometrical transformation from each element Ω^e to the reference element Ω^r.

For the sake of simplicity, we consider triangular elements and a linear approximation of $u(\mathbf{x})$ in each of those triangles. The element Ω^e will be defined from the coordinates of its three vertices $\mathbf{x}_1^e$, $\mathbf{x}_2^e$ and $\mathbf{x}_3^e$, with $\mathbf{x}_i^e = (x_i^e, y_i^e)$. We denote by ξ and η the coordinates in which the element of reference is defined. For linear approximations, the usual triangular reference element is defined by its three vertices $\boldsymbol{\xi}_1 = (\xi_1, \eta_1) = (0, 0)$, $\boldsymbol{\xi}_2 = (\xi_2, \eta_2) = (1, 0)$ and $\boldsymbol{\xi}_3 = (\xi_3, \eta_3) = (0, 1)$.

The linear interpolation associated to the three vertices is defined by

$$u^e(\xi, \eta) = N_1(\xi, \eta) \cdot u_1^e + N_2(\xi, \eta) \cdot u_2^e + N_3(\xi, \eta) \cdot u_3^e$$

$$= \left(N_1(\xi, \eta)\ N_2(\xi, \eta)\ N_3(\xi, \eta) \right) \cdot \begin{pmatrix} u_1^e \\ u_2^e \\ u_3^e \end{pmatrix} = \mathbf{N}^T \cdot \mathbf{U}^e, \qquad (A.18)$$

where $u_i = u(\xi_i, \eta_i)$ and the so-called shape functions $N_i(\xi, \eta)$ are given by

$$\begin{cases} N_1(\xi, \eta) = 1 - \xi - \eta \\ N_2(\xi, \eta) = \xi \\ N_3(\xi, \eta) = \eta \end{cases} . \qquad (A.19)$$

Notice that the approximation is linear, that it defines an interpolation and that the approximation on a triangle edge only depends on the function value at the two vertices that it joins, thus ensuring continuity of the approximation between adjacent elements.

The geometrical transformation from Ω^e to Ω^r can be expressed from

$$\begin{cases} x(\xi, \eta) = N_1(\xi, \eta) \cdot x_1^e + N_2(\xi, \eta) \cdot x_2^e + N_3(\xi, \eta) \cdot x_3^e \\ y(\xi, \eta) = N_1(\xi, \eta) \cdot y_1^e + N_2(\xi, \eta) \cdot y_2^e + N_3(\xi, \eta) \cdot y_3^e \end{cases} . \qquad (A.20)$$

Its Jacobian has in the present case an explicit expression:

$$\mathbf{J} = \begin{pmatrix} \frac{\partial x}{\partial \xi} & \frac{\partial y}{\partial \xi} \\ \frac{\partial x}{\partial \eta} & \frac{\partial y}{\partial \eta} \end{pmatrix} = \begin{pmatrix} -x_1^e + x_2^e & -y_1^e + y_2^e \\ -x_1^e + x_3^e & -y_1^e + y_3^e \end{pmatrix} . \qquad (A.21)$$

The Jacobian of the inverse transformation $\mathbf{Q}$ reads

$$\mathbf{Q} = \begin{pmatrix} \frac{\partial \xi}{\partial x} & \frac{\partial \eta}{\partial x} \\ \frac{\partial \xi}{\partial y} & \frac{\partial \eta}{\partial y} \end{pmatrix} = \begin{pmatrix} \frac{\partial x}{\partial \xi} & \frac{\partial y}{\partial \xi} \\ \frac{\partial x}{\partial \eta} & \frac{\partial y}{\partial \eta} \end{pmatrix}^{-1} = \mathbf{J}^{-1} . \qquad (A.22)$$

We can now come back to the weak form (A.17) and proceed to its discretization. For this purpose, we consider a generic integral of the left-hand-side of (A.17):

$$\int_{\Omega^e} (\nabla u^{e*})^T \cdot \mathbf{D} \cdot \nabla u^e \, d\mathbf{x}, \tag{A.23}$$

and apply the transformation $\Omega^e \rightarrow \Omega^r$ to obtain

$$\int_{\Omega^r} (\nabla_\xi u^{e*})^T \cdot \mathbf{Q}^T \cdot \mathbf{D} \cdot \mathbf{Q} \cdot \nabla_\xi u^e \cdot \det(\mathbf{J}) \, d\xi \cdot d\eta. \tag{A.24}$$

Here, we have considered that

$$\nabla u^e = \begin{pmatrix} \frac{\partial u^e}{\partial x} \\ \frac{\partial u^e}{\partial y} \end{pmatrix} = \begin{pmatrix} \frac{\partial \xi}{\partial x} & \frac{\partial \eta}{\partial x} \\ \frac{\partial \xi}{\partial y} & \frac{\partial \eta}{\partial y} \end{pmatrix} \cdot \begin{pmatrix} \frac{\partial u^e}{\partial \xi} \\ \frac{\partial u^e}{\partial \eta} \end{pmatrix} = \mathbf{Q} \cdot \nabla_\xi u^e. \tag{A.25}$$

Now, from the approximation (A.18), we can express $\nabla_\xi u$ as a function of its nodal values according to

$$\nabla_\xi u^e = \begin{pmatrix} \frac{\partial u^e}{\partial \xi} \\ \frac{\partial u^e}{\partial \eta} \end{pmatrix} = \begin{pmatrix} \frac{\partial N_1}{\partial \xi} & \frac{\partial N_2}{\partial \xi} & \frac{\partial N_3}{\partial \xi} \\ \frac{\partial N_1}{\partial \eta} & \frac{\partial N_2}{\partial \eta} & \frac{\partial N_3}{\partial \eta} \end{pmatrix} \cdot \begin{pmatrix} u_1^e \\ u_2^e \\ u_3^e \end{pmatrix}$$

$$= \begin{pmatrix} -1 & 1 & 0 \\ -1 & 0 & 1 \end{pmatrix} \cdot \begin{pmatrix} u_1^e \\ u_2^e \\ u_3^e \end{pmatrix} = \mathbf{B} \cdot \mathbf{U}^e. \tag{A.26}$$

This allows us to write (A.24) as

$$\int_{\Omega^r} \mathbf{U}^{e*T} \cdot \mathbf{B}^T \cdot \mathbf{Q}^T \cdot \mathbf{D} \cdot \mathbf{Q} \cdot \mathbf{B} \cdot \mathbf{U}^e \cdot \det(\mathbf{J}) \, d\xi \cdot d\eta. \tag{A.27}$$

Since $\mathbf{U}^e$ does not depend of the space coordinates, it can be moved outside the integral:

$$\mathbf{U}^{e*T} \cdot \left(\int_{\Omega^r} \mathbf{B}^T \cdot \mathbf{Q}^T \cdot \mathbf{D} \cdot \mathbf{Q} \cdot \mathbf{B} \cdot \det(\mathbf{J}) \, d\xi \cdot d\eta \right) \cdot \mathbf{U}^e = \mathbf{U}^{e*T} \cdot \mathbf{K}^e \cdot \mathbf{U}^e. \tag{A.28}$$

Considering a generic integral of the right-hand-side of (A.17), and the approximation of $f(\mathbf{x})$ in the reference element Ω^r according to

$$f^e(\xi, \eta) = N_1(\xi, \eta) \cdot f_1^e + N_2(\xi, \eta) \cdot f_2^e + N_3(\xi, \eta) \cdot f_3^e = \mathbf{N}^T \cdot \mathbf{F}^e, \tag{A.29}$$

with $\mathbf{F}^{eT} = (f_1^e, f_2^e, f_3^e)$, we obtain

$$\int_{\Omega^r} u^{e*} \cdot f \, d\mathbf{x} = \int_{\Omega^r} \mathbf{U}^{e*T} \cdot \mathbf{N} \cdot \mathbf{N}^T \cdot \mathbf{F}^e \cdot \det(\mathbf{J}) \, d\xi \cdot d\eta = \mathbf{U}^{e*T} \cdot \mathbf{M}^e \cdot \mathbf{F}^e = \mathbf{U}^{e*T} \cdot \mathbf{H}^e. \tag{A.30}$$

Thus, we can finally write (A.17) as

$$\sum_{e=1}^{e=E} \left(\mathbf{U}^{e*T} \cdot \mathbf{K}^e \cdot \mathbf{U}^e \right) = \sum_{e=1}^{e=E} \left(\mathbf{U}^{e*T} \cdot \mathbf{H}^e \right). \tag{A.31}$$

If we define $\mathbf{U}$ as the vector containing all nodal values of the approximation, that is, a vector of size N_n, the total number of nodes in the mesh, then (A.31) reads

$$\mathbf{U}^{*T} \cdot \mathbf{K} \cdot \mathbf{U} = \mathbf{U}^{*T} \cdot \mathbf{H}. \tag{A.32}$$

If vertice k ($k = 1, 2, 3$) of element e corresponds to node number $n(k, e)$ of the mesh, then entry (i, j) of matrix $\mathbf{K}^e$ must be added to entry $(n(i, e), n(j, e))$ of matrix $\mathbf{K}$, i.e. $\mathbf{K}_{n(i,e),n(j,e)} = \mathbf{K}_{n(i,e),n(j,e)} + \mathbf{K}^e_{i,j}$, for $i, j = 1, 2, 3$ and $e = 1, \ldots, E$. Analogously, entry i of vector $\mathbf{H}^e$ must be added to entry $n(i, e)$ of the global vector $\mathbf{H}$, i.e. $\mathbf{H}_{n(i,e)} = \mathbf{H}_{n(i,e)} + \mathbf{H}^e_i$, for $i = 1, 2, 3$ and $e = 1, \ldots, E$.

After assembling matrices $\mathbf{K}^e$, $\forall e$, into $\mathbf{K}$ and vectors $\mathbf{H}^e$, $\forall e$, into $\mathbf{H}$, (A.32) can be rewritten as:

$$\mathbf{U}^{*T} \cdot (\mathbf{K} \cdot \mathbf{U} - \mathbf{H}) = 0. \tag{A.33}$$

In view of the arbitrariness of $\mathbf{U}^*$, this implies

$$\mathbf{K} \cdot \mathbf{U} - \mathbf{H} = 0 \rightarrow \mathbf{K} \cdot \mathbf{U} = \mathbf{H}. \tag{A.34}$$

However, as indicated before, (A.16) applies for $u^*(\mathbf{x} \in \Gamma_D) = 0$. Thus, the rows in (A.34) related to nodes located on the Dirichlet boundary Γ_D cannot be enforced because the corresponding nodal value in $\mathbf{U}^*$ of (A.33) vanish. Thus, these rows must be substituted by the essential boundary conditions. It suffices, even if there are better alternatives, to put zeros in all the entries of rows of matrix $\mathbf{K}$ and vector $\mathbf{H}$ related to nodes located on Γ_D. Then by introducing a one in the diagonal entries in such rows of $\mathbf{K}$ and the given value of the solution at those nodes in the corresponding rows of $\mathbf{H}$, the resulting algebraic system can finally be solved to obtain the solution $\mathbf{U}$.